AF539373

Commercial Flower Cultivation

The Author

Professor (Dr.) Kaushal Kumar Misra received his Ph.D. Horticulture degree in the year 1986 from Govind Ballabh Pant University of Agriculture and Technology, Pantnagar-263145, U.K. He did his post doctorate degree from V. P. I. and State University, Blacksburg, Virginia, U.S.A. during 1989. He started his career as Senior Research Assistant on June 08, 1974 from Pantnagar. He was selected as Junior Research Officer, Horticulture on April 1, 1978. He was selected as Senior Research Officer / Associate Professor, Horticulture on April 1, 1991. Dr. Misra was promoted as Professor, Horticulture on July 27, 1998. He was Officer Incharge at Research Station, Sui, Lohaghat District Pithoragarh, U.P. from January 12, 1992 to March 31, 1994, then again as Associate Director/Joint Director at H.R.C., Patharchatta from November 19, 1997 to July 15, 1999. He was selected as Head of Department of Horticulture from November 1, 2004 to December 18, 2005 and September 30, 2006 to June 01, 2009. He published 21 books, 190 research papers, 150 popular articles and a dozon of book chapters in review books to his credits. He was awarded Dr. Rajendra Prasad Puruskar, 1984 from I.C.A.R., New Delhi, conferred, ISHRD Fellowship in September, 2004 from Indian Society of Horticultural Research and Development, Uttarakhand, Best Citizen of India Awards, 2006 and 2010, New Delhi and fellowship from Horticultural Society of India award, 2013. Sixteen Ph.D. and fourteen M.Sc. (Ag.) Horticulture students received his / her degree under his guidance. He has got about 41 years of experience of teaching, research and extension education.

Dr. Satish Chand is working as Junior Research Officer since 17 May, 2006 in G.B.Pant University of Agriculture and Technology, Pantnagar. He had completed Ph.D. in 2005 from H.N.B., Garhwal Center University, Srinagar, Uttarakhand. He has published 16 research papers, 25 popular articles, one book and one book chapter. He has guided 6 M.Sc. Horticulture (Floriculture and Landscaping) and one M.Sc. Ag. Horticulture (Pomology) students.

Commercial Flower Cultivation

- Authors -

Dr. Kaushal Kumar Misra
Ph.D., P.D. (U.S.A.),
Professor and Former Head

and

Dr. Satish Chand
Ph.D. (Horticulture)
Junior Research Officer (Horticulture),
Department of Horticulture, College of Agriculture,
Govind Ballabh Pant University of Agriculture and Technology,
Pantnagar-263145, District – Udham Singh Nagar, Uttarakhand

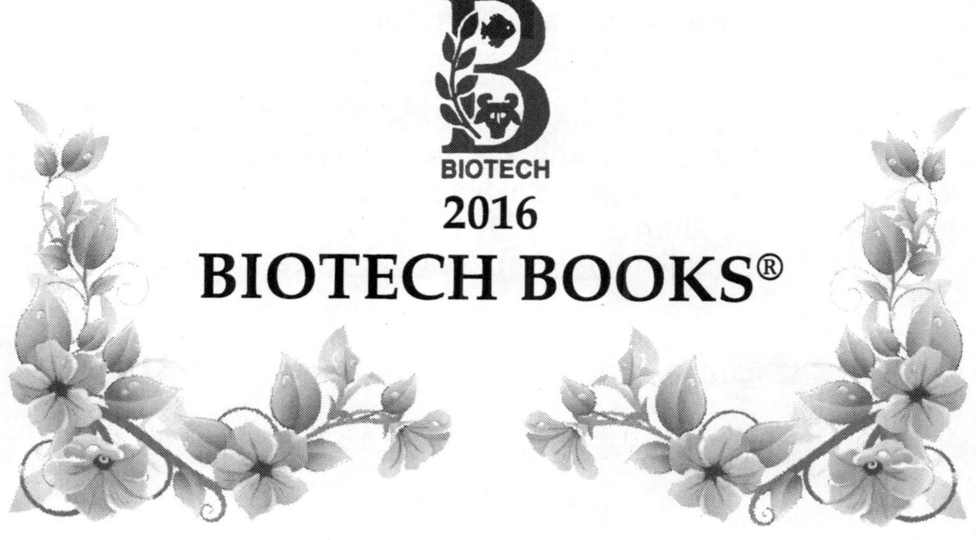

2016

BIOTECH BOOKS®

ISBN 978-81-7622-366-9

Published by: **BIOTECH BOOKS®**
4762-63/23, Ansari Road, Darya Ganj,
New Delhi - 110 002
Phone: +91-011-23262132
E-mail: biotechbooks@yahoo.co.in

Digitally Printed at : **Replika Press Pvt. Ltd.**

PRINTED IN INDIA

— Dedicated to —

My grandfathers

Shri Gaya Prasad Mishra
and
Shri Gaurishanker Pandey

Preface

"Commercial Flower Cultivation" has nineteen chapters. The first chapter deals with introduction. The second chapter deals with anthurium. The third chapter deals with bougainvilleas. The fourth chapter deals with calla lily. The fifth chapter deals with canna. The sixth chapter deals with carnation. The seventh chapter deals with chrysanthemum. The eighth chapter deals with dahlia. The nineth chapter deals with gerbera. The tenth chapter deals with gladiolus. The eleventh chapter deals with jasmine. The twelveth chapter consists of lilium. The thirteenth chapter consists of marigold. The fourteenth chapter consists of orchids. The fifteenth chapter consists of rose. The sixteenth chapter consists of tuberose. The seventeenth chapter consists of zinnia. The eighteenth chapter consists of floral ornaments. The nineteenth chapter consists of pest management. Each chapter has been divided into sub chapters which are very specific in nature.

Efforts have been made to describe various chapters in a systematic and comprehensive manner. The subject matter illustrated with figures and tables, wherever, felt necessary. The bibliography presented at the end can be useful to obtain further information. In the last glossary containing related terms have been presented. The book has been written keeping in view the requirements of the graduate, post graduate students, teachers and research scientists in the field of floriculture. The authors are grateful to several students at Govind Ballabh Pant University of Agriculture and Technology, Pantnagar. The first author is thankful to his wife Mrs. Manjul Mishra, daughter Mrs. Kanan Kaushal Pandey, sons Mr. Praduman Kaushal, Mr. Panini Kaushal, daughter-in-law Mrs. Pratibha Kaushal and Mrs. Neha Kaushal, son-in-law Mr. Satyajeet Pandey, nephew Mr. Naveen Kumar and grandsons Mr.

Pratham Kumar and Mr. Priyan Kaushal who helped in many ways in writing of this book. Thanks are also due to Mr. Shailendra Kumar, Senior Assistant for typing the manuscript on computer within the shortest possible time.

Dr. Kaushal Kumar Misra

Dr. Satish Chand

Contents

1

Introduction

The rapid expansion in production technology, breeding, physiology and tissue culture accompanied by plant protection and post harvest physiology have led great advances in research of the floricultural crops. Routine mass propagation methods have enormously increased the rates at which ornamental plant materials can be markeded. Novel methods of plant improvement involving gene transformation, protoplast fusion and recombinant DNA technology depend on tissue culture for regeneration of useful new plant materials like blue rose and carnation. The major ornamental plants where most of the advances in research have taken place are rose, chrysanthemum, carnation, gladiolus and orchids. Rose breeding has different objectives in different countries *e.g.*, inspite of many breeding efforts in Canada. There are very few cultivars that are sufficiently hardy to survive in winter and still be ever blooming. The species, *Rosa rugosa* which is extremely winter hardy, very vigorous, resistant to drought, mildew and blackspot have been crossed to chinensis group by *Felicitas svejda* in Ottawa has resulted into hybrids resistant to these diseases besides, winter hardiness. In carnation, *Dianthus spp.* which vary in ploidy level being diploid to hexaploid there has been the constant desire for improving winter hardiness combind with flower garden carnation, advances in breeding programme utilizing 20 species and 12 garden cultivars produced 25000 seedling resulting into various diploid (Pink Princess), Cheyenne (5X) and War Bonnet (4X). Most commercial green house carnations are induced sports of cultivar William Simone of the standard type which provides a very narrow genetic base and a deficiency of dark pink and crimson and blue flower colours. F_1 seed production in auto tetraploid and cultivars as War Bonnet and Cheyenne give little seeds. Back-crossing to the antatetraploids improved quality but seed set was lost by third backcross generation. However, intercrosses of BC_2 and BC_3 generation maintained fertility with newer and novel forms.

Improved cultivated and green house Chrysanthemums are products of over past 2000 years breeding programme. They enjoy popularity in the international market and some more than 6000 cvs. are available, Breeding protocols have been improved and optimal conditions for light, temperature, nutrition and pest control determined. *In vitro* mutagenesis has proved useful for and bioreactor somatic embryoid production may amend asexual propagation and breeding. These technological innovations bypass sexual cycles and germplasm improvement involving genetic recombination. Sexual breeding programmes and full of difficulties for the production of homozygous inbred lines required for hybrid seed production or germplasm for genetic studies. The technique of embryo rescue has been successfully utilised in many crops to reduce the generation time and circumvent post fertilization barriers in order to recover interspecific hybrids. It is also a well developed method for orchids where the seeds are without endosperm. Radiation induced mutants have been developed by determining the nature of radiation induced mutations and associated physiological and morphological changes and their propagation by asexual means. New varieties of mums have thus originated for green house bench, plastic 10 cm. pot culture and also for outdoor culture which are early bloomers and seed propagators.

In recent years, gladiolus has become one of the prized commercial cut flower crop. Breeding of newer and novel forms and colours on long and synchronous flowing stalks, the number of florets per spike is another important character. Florida Flame and Sylvia are resistant from *Fusarium oxysporum* var. gladiorum. Production of fragrant gladioli by corssing *Acidenthera murielaeith* cv. Filigiri name as 'Lucky star' has opened new vistas in the production of scented gladiolus. Vigorous growth of plants recurrent flowering habit, exquisite flower form, shape, size, floriferousness, witner hardiness, resistance to diseases and flowers having nice fragrance have rose breeding objectives. Long stem, slow opening and long pointed bud for the cutflowers have different characters. There are two broad classes now a days, hybrid tea and floribunda with miniatures and climbing types. The cvs like 'Bhim', 'Priyadarshini', 'Pusa Sonra', 'Navneet' and other are resistant to powdery mildew under field conditions.

Flowers and plants have become an integral part of human living. They are useful in improving the quality of life. Nowadays, ornamental plants play a significant role in environmental planning of urban and rural areas, social and rural forestry, waste and development, afforestation and landscaping of outdoor and indoor spaces. Floriculture is also economically important. It generates gainful employment of youths in sub-urban and rural areas, especially in plant nursery business and hybrid seed production. The flower crops provide higher income from comparatively small areas with high profitability as compared to other crops. However, the advanced floriculture technology is capital intensive in view of high cost of greenhouses, equipments, machines, chemicals and other infrastructural facilities. There is a wide choice of crops, species and cultivars suitable for growing successfully in different climates and regions of the country. The ornamental plants and cut flowers are non-traditional items with high potential for export. They are useful in beautifying outdoor and indoor spaces. Plants are useful in amelioration of environment by overcoming pollution to a great extent. Floriculture provides both profit and pleasure to all engaged in it. It improves the quality of life. Florist trade of cut flowers and cut foliage which may fresh, dried, dyed, bleached, impregnated, otherwise prepared in the form of garlands, bouquets, floral baskets, floral ornaments and flower decorations. It may be either traditional using stalkless flowers and the value added products like garlands, veni, floral ornaments and decorations, petals or contemporary trade of long-stemmed flowers, bouquets, floral baskets and flower arrangements. Plant nursery for propagation and supply of life plants including those multiplied by tissue-culture. Bedding Plant industry for supply of seedings and rooted cuttings of flowers, like Chrysanthemum, carnation, gerbera, dahlia, poinsettia and many other flowers. In India the bedding plant business has not been developed as in the U.S.A., the Netherlands and other European countries. Production and sale of seeds and bulbs of flower crops including hybrid seed production. Flower perfumes especially for production of essential oil, attar, concrete which are required by cosmetic, food and flavour industries. The two most important flowers for perfume production are rose and jasmine, while some others like tuberose, lavender and geranium are also used for this purpose.

Users of Floricultural Products

Professional gardeners who tend the gardens and parks, both public and private. Commercial flower growers including large traditional class of flower growers called the Mali. These Malis have small and marginal holdings near towns whose livelihood depends exclusively on growing of flowers for garlands, bouquets, floral ornaments and floral decorations required on weddings, religious and other ceremonies and social occasions, Unfortunately they have remained economically weak and there is almost no change in their cultivation practices. There has been no effort made to improve their technology and economic conditions. Landscape architects, consultants and contractors utilize live plants and flower plants in their landscaping projects. In big cities, there are plant rental services available for indoor landscaping and maintenance of plants in commercial complexes, offices, banks and other business centres. For public parks and gardens, the town planners, public horticulture departments and municipal corporations have regular demands for ornamental plants. Cut flowers and foliage of various kinds are required by the florists and small flower traders. Amateur home gardeners, hobbyists and housewives are also common users of flowers and plants. Live plants, cut flowers and foliage, seeds and bulbs are being exported on a small scale but their demand is likely increase in future.

Subservient Industries

Steel, aluminium, glass, fiberglass, polythene, plastic and polycarbonate required for greenhouses. For packaging of flowers and plants cardboard, paper, polythene, plastic are commonly used. Cut flowers and live plants are transported in vans or other vehicles. In India bicycles, cycle-rikshaw, autorikshaw are also commonly used for transportation of flowers and plants and in some cases these are send to distant markets by aeroplanes. Cut flowers are generally hardened in cool storage rooms before packing in cardboard boxes. These are also transported in cool vans. In the modern florist shops there are airconditioned rooms. Horticultural inputs like, pesticides, fertilizers, chemicals, tools, implements, equipments, machines, pots and other containers are in great demand in floriculture. Many scientific apparatus, equipments and chemicals are required by the tissue-culture laboratories which have undertaken large scale multiplication of ornamental plants.

Agrotechniques and Quality of Products

In general, both agrotechniques and quality of the products are poor and inconsistent. The varieties being cultivated presently are in many cases old and obsolete, low yielding and inferior in quality as compared as the modern cultivars grown in other countries having advanced floriculture, like Holland, USA, U.K. and France. Planting material, including seeds, bulbs, cuttings of desired varieties is either not available or the quantity available is insufficient and of sub-standard in many cases. Besides, pre and post harvest losses at various levels in cultivation and marketing of flowers are high mainly due to lack of proper infrastructural facilities, like greenhouses, equipment, cool storage and transportation facilities. In India recently the rise in growth of floriculture has primarily depended upon three factors, such as, changes in social values of people, increases in income-levels, especially disposable income levels and changes in urban-rural population mix and population increase in cities. The social values of the people are undergoing extensive changes due to the progressive urbanization of the economy and the transformation in the living style. The rural income level is expected to grow at 1.5 per cent annually. The differential in urban-rural income levels has been projected at 3 per cent. Thus urban income is likely to increase at a rate of 4.5 to 5 per cent. The population growth rates have been projected about 2 per cent annually. However, there are much higher for the metropolitan cities like, Delhi (5 per cent) and Bangalore (6 per cent) resulting in the increase of the urban-rural population mix contributing to increased demand in floricultural products. The tourism and hotel industry can be considered a useful parameter to quantify the changes in the social values. The tourism was expected to grow to 3.5 million tourists in 1990 and 18 million by 2280 A.D. Efforts would be planned to increase the hotel industry so that the targeted tourist levels may be achieved. During the period 1963-2032 the number of tourists have increased from 141 thousand to 363 thousand, *i.e.*, by 510 per cent, whereas, the hotel rooms have increased from 7 thousand to 22 thousand during the care puries by 814 per cent. Taking into consideration the socio-economic changes, a growth rate of 20-25 per cent in floriculture industry appears to be a reasonable estimate for the future period.

There is an increasing demand for floricultural products in the world. Cut flowers and live plants are the important items of import and export in the present day international trade. The consumption of flowers and plants per capita is the highest in Switzerland (64 US dollars) followed by Holland (60 dollars) and Germany (58 dollars). However, in view of the national income per capita in these countries, the consumption is considered to be low and it is expected to continue to rise in future. The production of cut flowers and plants in most of the European and North American countries cannot meet the domestic demand and therefore, their requirement is to be met by the imports. In all the European markets, the demand for cut flowers is at peak during winter months (November to March), particularly for the Christmas period, Valentine's Day, Mother's Day, All Saint's Day and Easter. In the cold season when the production is in green-houses, the limitation of space and high cost of labour and energy result in limited quantity and higher cost of production as compared to that in the summer. A few developing countries of the tropical and sub-tropical areas have taken up exports of flowers during winter when it is possible for them to produce these at a comparatively lower cost of production than in Europe. But unfortunately the share of developing countries in the world exports is very low (15.4 per cent). Germany is the biggest consumer of cut flowers and the Netherlands ranks the highest in exports, its share being as high as 66 per cent in cut flowers and 45 per cent in plants.

Among the three cut flowers in the international imports, carnation ranks first followed by rose and chrysanthemum. The Netherlands is the leading exporter of roses followed by Colombia and Israel. The four largest exporters of carnation are Colombia, Netherlands, Israel and Italy. Chrysanthemums are supplied mainly by Netherlands and Colombia. Colombia sends its supplies mostly to U.S.A. and some to European countries. The maximum imports is by Germany, about 86 per cent of the total world imports, while the U.S.A., France, United Kingdom, Switzerland and Netherlands are the other important importing countries and its exports include rose, carnation, chrysanthemum, alstcemsria and statice. The other developing countries which supply cut flowers are Taiwan, Singapore, Peru, Mexico, Costa Rica, Brazil, Ethiopia, Zimbabwe, Mauritius and Malaysia. Both Singapore and Malaysia export orchids and Mauritius supplies anthurium. The South American countries (Peru, Mexico, Costa Rica and Brazil) export rose, carnation, chrysanthemum and other flowers. The Netherlands obtains roses and gladiolus from Zimbabwe during winter while gladiolus is also provided by Ethiopia in winter. While analyzing the market opportunities of cut flowers in foreign markets, it has been observed that carnation, orchids and chrysanthemum from developing countries may have favourable markets in Europe. France is a good market for exporting rose, spray carnation, chrysanthemum in winter and orchids, particularly new varieties and species. The long distance shipment of cut flowers to USA from Asian and African countries may not be competitive with supplies from South American region.

Live Plants

The world imports and exports include a large variety of plants, both ornamental foliage and flowering. Among the foliage plants, the most important ones are Dieffenbachia, Dracaena, Ficus, Philodendron, Algaonema, Marants, Croton, Yuosa, Pegonia, Sainpaulia, Kalandhoe Palargonium, Corysyline, Scindapsus, Syngoniu, Ananas, Sonefflera, Bromeliage, Palms, Spathiphyllum. The total world imports and exports of live plants are 1035.29 m and 989.31 m US dollars, respectively. The largest buyer is the Germany (23.6 per cent) followed by France, United Kingdom, Italy and the Netherlands. The leading exporter (45.3 per cent) is the Netherlands and the other biggest exporters are Denmark, Belgium, Luxembourg, Germany, France and U.S.A. The live plants exported to Europe include rooted and unrooted cuttings, large 'finished' plants and 'semi-finished' plants. The terminal unrooted cuttings with 5 to 6 leaves of *Ficus elastica* are supplied in the European markets. The 'finished plants' of *Ficus bengiamina* and others are exported mainly from California and Florida by sea. However, these require long period of nine months of acclimatization in the importing country because most of the leaves shed during transportation. Large scale propagation of foliage plants by tissue culture has been taken up in the laboratories established in the Netherlands and Belgium. The tissue-cultured plants with better rooting, superior quality, uniformity and disease free have proven to be advantageous in spite of higher price as compared to those propagated by conventional methods. The low-wages economy and low cost of propagation in the developing countries have been offset by the consistent and timely supplies of quality plants propagated by tissue-culture available from the nearby countries, like Holland and Belgium. In view of these recent developments, the market opportunity in European countries has become limited. Never the less, there is still a demand for such plants like Dracaena and *Ficus alastica*. The propagation of which by tissue-culture does not have much advantage. The export of live plants from the developing countries can be viable, if the air-freight is low and the quality and supplies are consistent.

Export from India

The first trial consignments of roses (cut flowers are sent to Paris, Amsterdam and Frankfurt markets in 1969 by the floriculture scientists of the Indian Agricultural Research Institute with the assistance of the State Trading Corporation of India. This successful venture paved the way for the export of floricultural products from the country. Unfortunately there has not been any significant progress in the export trade of ornamentals. The export of floricultural products from India is almost negligible. It was Rs. 84 lakhs in 1983-86. Almost half of the quantity exported comprised of foliage plants. However, during this period, some encouragement and support came from the Government of India agencies, like the Ministry of Commerce, Agricultural and Processed Food Products Export Development Authority, State Trading Corporation and Trade Development Authority. A trade delegation of growers, trade and marketing sponsored by the Ministry of Commerce visited Moscow, Amsterdam, Hamburg and Frankfurt in 1982 to study the various aspects of export of flowers and live plants which later submitted its report alongwith the recommendations to the Government of India.

Intensive Floriculture Areas

The following intensive floriculture areas have been recommended by Export Group on Floriculture Development in India.

Constraints in Export

The following have been identified as the constraints and limitations in the development of an export based floriculture industry in the country. Non-availability of germplasm of important flowers and plants which have an export market. Lack of sufficient knowledge or knowhow on high-tech production practices including important methods of propagation such as mist-houses for micro-propagation. Lack

Table 1.1: Intensive Floriculture Areas of India

Sl.No.	Areas	Crops
1.	Area around Delhi	Rose, Gladiolus, Carnation, Crysanthemum.
2 and 3.	Area around Pune/Nasik	Rose, Carnation, Gladiolus, Dahlia, Chrysanthemum.
4.	Area around Bangalore	Rose, Gladiolus, Carnation, Chrysanthemum and other Ornamental Foliage Plants.
5.	Area around Trivendrum	Orchids, Anthurium, Foliage plants.
6.	Area around Kalimpong	Orchids, Amaryllis, Gladiolus, Lilium, Gerbera, Anthurium, Cacti and Foliage Plants.
7.	Area around Calcutta	Lotus, Tuberose, Jasmine, Chrysanthemum, Dahlia, other Ornamental/Foliage Plants.
8.	Area around Srinagar	Gladiolus, Lilium, Carnation, Rose, other Bulbous Plants, Seeds
9.	Area around Solan	Gladiolus, other Bulbous Plants and Seeds.
10.	Area around Coimbatore including Nilgiris	Jasmine, Tuberose, Chrysanthemum, Rose, Carnation, Orchids.

of training and extension facilities for transfer of even existing knowhow. Non-availability of suitable organizations in the private, cooperative and public sector who can take up the capital intensive activities of production of planting material and assured marketing facilities. Lack of adequate post-harvest infrastructure covering the areas of refrigerated transport, storage and packing. Ignorance of world trends in floricutlrue trade, international practices including patent rights as also varietal choice and demand. Cumbersome procedures for clearance of imported planted materials by customs and quarantine authorities. High rates of customs and import duties inhibiting import of input materials including glasshouse components, plant media, plant protection chemicals and fertilizers. Lack of sufficient incentives for export oriented floricultural units. Inadequate availability of cargo space and irrational air freight structures in comparison to rates available in competing countries such as Srilanka, Thailand, Pakistan and Kenya.

Suggestions

Availability of planting material/seeds of the preferred varieties in the international trade. Smooth and efficient quarantine facilities. Waiver of import duties on all inputs including planting material and equipments relating to floriculture export units. Establishment of integrated tissue culture laboratory and glass/greenhouse with foreign collaboration having buy-back arrangements. Establishment of Horticultural Estates for the cultivation of Floricultural crops. Formulation of schemes with adequate incentives to encourage farmers to take up flower production. Provision of extension and training facilities to potential flower growers. Post-harvest handling knowhow should be made available to growers. Appointment of trade delegations for international market study. Availability of refrigerated transport from the production points to airports. Availability of cold storages at the airports.

Availability of low interest rate loan from the nationalized banks. Lower air-freight charges.

Conclusion

There is an immense potential for export of floricultural products. The present situation of the foreign markets indicates that there is a good score for export for carnation, chrysanthemum and orchids preferably to France during winter months. Roses grown in greenhouse or under partial cover may have a possibility to be exported provided the quality is acceptable to the foreign buyers and the supplies are consistant. Although good quality gladiolus flowers can be produced in different parts of India, its export may not be competitive because of its high weight-to-volume ratio and its transportation in an upright position. Efforts should be made to develop export trade of orchid plants and cut flowers at Kalimoong, Darjeeling, Sikkim and Arunachal Pradesh for temperate orchids (Cymbidiums, Paphiopedilums, Vanda etc.) and in Trivandrum, Cocchin, Western Ghats, near Yersawd and similar areas for tropical orchids (Dendrosiums, Arands, Arachnis, Aranthers, Ondidium). It would be necessary to create adequate facilities for tissue-culture laboratories, greenhouses and introduction of important hybrids of the export markets and their rapid propagation. An intensive programme of orchid breeding may be undertaken with a view to developing new attractive hybrids, utilizing some of the native Indian Orchid species. Many beautiful hybrids have been evolved abroad by hybridization of the Indian orchid species with other species and genera.

The export of live plants, particularly foliage plants can be developed in Bangalore, Pune, Calcutta, Trivandrum and Cochin. But this sector of the floriculture industry needs modernization by having facilities of environment-controlled greenhouse, seran covered houses, tissue-culture laboratories, modern equipments for steam sterilization of soil-pot mix, plant protection and mist propagation. The availability of peat mix and inexpensive plastic pots and packaging materials are basic necessities in the export of live plants. While concluding the following suggestions are offered. Emphasis should be laid on developing export trade of cut flowers of orchids, carnation, chrysanthemum and roses at ideal locations, preferably near/national airport. Growing techniques of orchids and roses for cut flowers in glasshouses may be standardized. Rose, chrysanthemum and carnation may also be grown under partial cover for producing good quality flowers. "Cool-chain" must be provided from farm to packing house to airport. Reduction in present air-freight structure is necessary for the export of cut flowers and live plants to be competitive. The export formalities, like plant quarantine, custom clearance, certificates from forest department and CITIES authorities (for orchids) may be simplified and made expeditions. Financial institutions should provide loans on low rate of interest and easy terms of repayment. Import of equipments for greenhouses and tissue-culture laboratories may be liberalized. The introduction and propagation of planting materials for distribution to growers should be centralized in main growing areas. For accelerating the growth of the floriculture industry, three types of interrelated agencies have been suggested. The developmental agency will be related to the activities of research institutions, market intelligence for export, experimental/

demonstration farms, Centres of training and education and horticultural/ floricultural societies. The Regulatory Agency is necessary for handling the matters certaining to air-lines, plant quarantine, import, export and financing. The servicing agency may be closely linked with the Developmental and Regulatory Agencies and it will provide assistance to the farmers, cooperatives of farmers or big business/ export houses.

2

Anthurium

The anthurium is a perennial herbaceous plant usually cultivated for its attractive, long lasting flowers. *Anthurium,* for its exquisiteness, durability and by the terms of long vase-life stands out among most of the tropical cultivated flowers. As a member of the Araceae family, *Anthurium* is native to the tropics of Central and South America. Anthuriums are believed to be hybrids of *Anthurium andraeanum* Linden ex André with several closely related species in the section *Calomystrium* and have been referred to as *Anthurium andraeanum* Hort. Anthurium is commonly known as Tail flower, flamingo flower, Hawaiian love plant, cresta de gallo or tongue of fire. Anthurium has nearly 1,000 species, making it the largest genus in the plant family Araceae. The genus name "Anthurium" is from the Greek "anthos" (flower) and "oura" (tail), referring to the slender tail-like spadix that protrudes from the spathe. The specific epithet (species name) "andraeanum" was given in honor of Edouard Francis André, a French botanist, horticulturist, horticulture professor and editor, and landscape architect, who collected the species in Colombia in 1875.

Origin

All species of Anthuriums are native to the tropical rain forests of Central and South America, with *Anthurium andraeanum* indigenous to Colombia. The genus *Anthurium* consists of some 500-600 known species. Some of important cultivated species belonging to both flowering and foliage groups are:

Flowering Group

Anthurium andreanum, Anthurium bakeri, Anthurium brownie, Anthurium ornatum, Anthurium robustum and *Anthurium regale.*

Foliage Group

Anthurium clarinervium, Anthurium corrugatum, Anthurium crystallinum, Anthurium holtonianum, Anthurium leuconerum and *Anthurium panduratum.*

Climate

Anthurium is a tropical plant and therefore prefer average to warm interior environments, 65°C and 80°C. Anthurium grow best when there is little difference between day time and night time temperatures. During winter dormancy periods, they prefer lower temperatures *i.e.,* 60°C to 65°C. It also prefer moderate to high humidity levels (70 - 80 per cent), so mist leaves frequently with lukewarm water, place pots on a pebble tray or place a humidifier in the room.

Cultivars

- ☆ **Dutch Cultivars:** These generally require cooler conditions than other types.
- ☆ **Anneke 141**: Pink flower - rapid growth, high production.
- ☆ **Cuba**: Large white flower, high production.
- ☆ **Claudia 108**: Large red flower, moderate branching.
- ☆ **Lydia 420**: Pink splash, rapid growth, high production.

Hawaiian Cultivars

- ☆ **Ozaki** - Light red, broad, heart-shaped spathe with reddish purple erect spadix; high yield of flowers and suckers; requires more shade than other cultivars, susceptible to anthracnose.
- ☆ **Kaumana** - Dark red, small to medium spathe with reclining white spadix; high yielding, quick growing with many suckers; highly susceptible to anthracnose.
- ☆ **Kansako**- Medium-sized red spathe with overlapping lobes; high yielding with long stems which hold flowers high, protecting them from bruising by foliage.
- ☆ **Nitta**- Broad, orange, heart-shaped spathe with overlapping lobes and white reclining spadix; vigorous, high yielding with many suckers; tendency to produce flowers parallel to the stem.

- ☆ **De Weese**- Small, white open heart-shaped spathe with yellow reclining spadix; fairly high flower yield and large number of suckers.
- ☆ **Marian Seefurth**- Broad, rose-pink heart-shaped spathe with unusually large overlapping lobes and greenish-yellow spadix which droops when young and becomes erect when mature; exceptionally high yielding and suckers freely; highly resistant to anthracnose.
- ☆ **Obake types:** These are extremely variable in size and shape and usually exhibit some chlorophyll in the spathe. They are becoming increasingly popular. Cultivars include Anuenue, Chameleon and Mauna Kea.
- ☆ **Novelty Types:** These include the tulip types which have an upright and cupped spathe.
- ☆ **Calyps** - Dark pink in colour.
- ☆ **Trinidad** - Off-white colour with a maroon flush.

Growing Media

Anthuriums grow best in a well-aerated medium with good water retention capability and pH of 4.5-5.0 with good drainage. A good medium needs to be able to anchor the roots and stem so that the plant will not topple over as it grows larger, yet provide sufficient moisture, nutrients and aeration to the plant. Organic matter (*i.e.*, wood shavings, sugar cane bagasse, tree fern chips, taro peel, macadamia nut shells or coffee parchment), volcanic cinder, or an artificial medium (*i.e.*, rockwool, polyfenol foam) can serve as a good medium to anchor roots for anthurium plant growth and flower production. Volcanic cinder is presently the most utilized medium among commercial growers because of its availability and relatively low cost. Many growers also utilize a cinder/organic media mix for moisture and nutrient retention.

Propagation

Anthurium is propagated by seed, sucker and tissue culture.

Seeds

To grow Anthurium plants from seed is a lengthy process. It may take 3 years from seed to bloom. Propagation by seed is not recommended as a commercial propagation method as it results in high variability.

Suckers

Commercial propagation for the cutflower market is usually vegetative. Mature plants produce "aerial shoots" or "suckers" which will flower 3-6 months after planting.

Tissue Culture

More recently, the tissue culture technique has been used to rapidly propagate plantlets from promising specimens and these *in vitro* plantlets have become available to commercial growers. The CARDI tissue culture laboratory has been successful in the micro-propagation of the Caribbean Pink and orange tulip types. The plants obtained from culture must be potted and "hardened" under mist before being planted in their final positions.

Planting

Planting distance varies with the vigor of the cultivar, the type of shade and the planting rotation plan of the grower. Generally, 'Kaumana' may be planted closer than 'Kozohara', which produces larger flowers. Saran cloth houses can accommodate denser planting than that of plants grown under tree ferns. Hard-covered houses can accommodate even higher densities because of improved disease control. The most common bed planting distance is 30 cm x 30 cm plantings providing approximately 25,000 plants per hectare. Dense plantings decrease air circulation and hinder chemical spray penetration, hence it encourages disease and insects damage on anthurium. Anthuriums can also be grown in containers such as plastic pots or bags. For 20 cms. plastic pots, approximately 40,000 to 45,000 plants per hectare can be cultivated.

Manuring and Fertilization

The media into which anthuriums are planted provides very little nutrients for the plant. It is important that anthuriums be supplied with moderate but consistent levels of complete fertilizers. Generally fertilizers are absorbed through a plants roots and leaves. However, Anthurium leaves contain a thick layer of wax and thus do not absorb fertilizer particularly well. It is for this reason that most large commercial Anthurium operations fertilizer via their trickle irrigation system.

The three main elements in fertilizer are:

Nitrogen (N) for healthy green leaf growth. N is required in the greatest amount.

Phosphorous (P) for strong roots and stems.

Potash (potassium) (K) (for fruiting and flowering).

A typical soluble (dissolves in water) NPK fertiliser for anthuriums has the following analysis: 12:5:15 which means it contains 12 per cent nitrogen, 5 per cent phosphorous and 15 per cent potassium.

Table 2.1: Nutrient Deficiency Symptoms on Anthurium Plant

Elements	*Symptoms*
Nitrogen (N)	Necrosis (dead tissue) spots on the leaves and a yellowing of the old leaves.
Phosphorous (P)	The edges of the old leaves turn yellow. The young leaves have a hard, dark green colour and are much smaller than the younger leaves.
Potassium (K)	The old leaves look chlorotic (lose their normal green colour) among the veins while the general leaf colour turn light green. The young leaf is smaller with a reddish or dark green colour.
Calcium (Ca)	Young leaves as irregular, chlorotic spots. The leaves become pointed in shape.
Magnesium (Mg)	Yellowing of old leaves along the main veins. The veins often remain green in colour.

Irrigation

Irrigation is needed for optimum production, especially during dry periods, if anthuriums are grown in a porous or well-drained medium. To minimize spread of disease, surface irrigation is best.

Leaf Pruning

Leaf pruning is must be rigidly followed to keep disease and insect damage at a minimum. An anthurium plant may be pruned to a minimum of four leaves without any adverse effect on flower production and quality.

Mulching

Coconut husks, bagasse or coconut fibre bast is used to mulch plants after planting and approximately twice per year thereafter.

Replanting

Replanting or thinning of plants is necessary every 4-5 years to avoid reduction in blooming rate. Earlier thinning may be necessary if large side shoots are used as planting material.

Harvesting

Anthurium is usually harvested once per week. After harvest, the flower stem should be placed into a clean bucket filled with 10 to 15cms. of clean tap water. The water in the buckets should be changed regularly, at least every day after 2 days of use. The maturity of the flowers for harvesting is determined by firmness of the peduncle and the degree of colour change of the spadix. The botanical flowers mature on the spadix from the base towards the apex. As they mature, a change in colour can be discerned that moves from the base to the tip of the spadix within a period of 3 to 4 weeks. After the lower half of the spadix has changed colour, the anthurium is referred to as one-half mature. Flower maturity affects postharvest longevity. Fully mature flowers last 10 per cent longer than half-open flowers. Maximum vase life occurs when anthuriums are harvested with percentage or more of the flowers on the spadix open. Anthurium flowers should be harvested at this stage. Harvesting is best done after 4 p.m.

Post-harvest Handling

Anthurium vase life is probably limited by plugging of the stem water conducting tissue. There is conflicting evidence as to whether it is bacterial or physiological blockage of the water conducting tissue that reduces anthurium vase life. Both causes can be reduced by proper postharvest handling. The use of clean buckets and clean water to hold cut flowers are an essential first step. Two postharvest treatments have been found to increase postharvest life:

1. Treating the recently cut stem with 1000 ppm silver nitrate for 10 to 20 minutes.
2. Waxing the flower with wax formulations. However, only a few wax formulations are effective.

Ethylene in a low molecular weight gas that is notorious for reducing the quality of floral crops. In anthuriums, it causes spathe blueing and shedding. The vase life of anthuriums can be cut almost in half by a short exposure to ethylene. Ethylene sources include ripening fruit, diseased and injured plants and internal combustion engines. Exposure to ethylene should be avoided. Anthuriums should be packed carefully to reduce the risk of injury during shipping. Mechanical injuries due to creasing of the spathe or bruising from the spadix touching the spathe during shipping are major problems. These injuries lead to unsightly black marks and downgrading of flowers. Anthurium flowers should not be held in the packed condition during shipping for more than four days. During shipping, temperature extremes can cause loss of flowers. Flowers should not be exposed to temperatures less than 50°C for more than 1 day.

Chilling injury occurs below these temperatures. In anthuriums, it is expressed as darkening of the flower spathe and spadix. Commercial preservative should be recommended for the retailer and consumer. These solutions assist in reducing microbial growth and therefore, contribute to the maintenance of water uptake. Submerging the flowers in water or misting regularly to revive wilted flowers are also useful recommendations.

Grading

Damaged flowers must be culled. The undamaged flowers are graded according to size. International standards are based on Hawaii's standards but the market requirements depend on the importing country.

Table 2.2: Hawaii State Department of Agriculture Standards for Anthurium

Grade	*Average of Length Plus Width of Spathe*
Miniature	Under 7.5 cm (7.5 cms.)
Small	Over 7.5-10 cm (7.5-10 cms.)
Medium	Over 10-12.5 cm (10-12.5 cms.)
Large	Over 12.5-15 cm (12.5-15 cms.)
Extra Large	Over 15 cm (15 cms.)

Table 2.3: The Grades have been Established in Jamaica

Grade	*Average of Length Plus Width of Spathe*
Miniature	5-7.56 cm (5-7.5 cms.)
Small	Over 7.5-10 cm (7.5-10 cms.)
Medium	Over 10-12.5 cm (10-12.5 cms.)
Large	Over 12.5-15 cm (12.5-15 cms.)
Extra Large	Over 15 cm (15 cms. inch)
Premium	Superior large blooms with 45.7cm (45 cms.) straight stems, flat spathe and inclined spadix.

Packing

For export, blooms are usually packed in cardboard cartons. Carton size will depend on the requirements of the individual importer but examples are given below:

Table 2.4: The Carton Dimensions of various Grades

Grades	*Carton Dimension*		
	Length (mm)	*Width (mm)*	*Height (mm)*
Small	964	137	65
Medium	964	190	65
Large	964	295	65

Each flower should be wrapped individually in a damp paper sleeve to protect against damage during transport. Shredded paper without print should be used for cushioning the blooms. Cartons should be lined with damp paper which should fold over the flowers. Sometimes, a strip of foam and a small piece of wood are fitted across the width of the box, mid-way along the stem length to keep the stems in place during transport. Boxes must be properly stapled and not too moist. They must never be left in the sun. Blooms may be transported to local outlets in buckets of clean water. It is recommended that flowers be tied in bunches of one dozen and a sleeve placed on each bunch to protect it from mechanical damage.

Storage

Blooms should be stored at temperatures between 13 and 17°C. They often remain in excellent condition in water for 3-4 weeks at 13°C. Storage at 7°C and below will cause chilling injury resulting in darkening of blooms. The use of preservatives such as silver nitrate and silver thiosulphate solutions may extend the vase life of some cultivars four times over that of flowers kept in water.

Insects and Nematodes

Table 2.5: Insects and Nematodes of Anthurium

Names	*Symptoms*	*Recommended Control*
Army worm, Aphids, Thrips	Holes in young developing leaves. Brown scars on spadix	Use Dipel or pyrethroids like Sherpa', or Orthene' at 5g/10 L of water
Snails and slugs	Brown scars on spadix	Metaldehyde bait applied to the surface of the growing medium.
Mealy bugs, Scale insects	Loss in plant vigour. Growth of sooty mould fungus on leaves and flowers.	Malathion' at 100ml/10L water or Diazinon' at 26m1/10 L water.
Grasshoppers	Holes in leaves and spathes	–
White flies	Loss of plant vigour and decreased flower production	Diazinon at 26 m1/10 L.
Nematodes	High populations may cause yellowing and falling of leaves.	Apply carbofuran (Furadan) to soil at 3.4kg/1000m^2.

Diseases

Table 2.6: Various Diseases of Anthurium

Common Name	Causal Organism	Symptoms	Recommended Control Methods
Root Rot	*Pythium* spp. or *Phytophthora* spp.	Rotted roots, fewer smaller leaves and flowers	Improve drainage and aeration. Drench with Banrot at 3.7-7.5g/10 L water or Captan at 22g/10 L water or Thiram 11-22g/10 L water.
Corticium fungus	*Corticium* spp.	White mycelial growth on surface of leaf sheath base of plant	Apply Kocide 606 at manufacturers's recommended rate.
Anthracnose	*Colletotrichum gloeosporioides*	Circular to triangular black spots on petiole blade, spathe and spadix	Alternate sprays of benomyl (Benlate) at 11g/10 L water with Dithane M45 at 22g/10 L and Maneb at 22g/10 L water.
Bacterial blight	*Xanthomonas campestris*	Deformed flowers with water-soaked areas from top downwards.	Drench with Ridomil, Dithane M45 or Aliette at manufacturers' recommended rates.
Mosaic	virus	Deformed mottled leaves. Blooms distorted and unsaleable.	Remove affected plants.

Physiological Disorders

Table 2.7: Physiological Disorders of Anthurium

Disorders	Symptoms	Recommended Control
Decline	Reduced plant and flower size as well as flower number	Raise beds to improve drainage. Replenish growth medium. Control nematodes.
Aborted spadix	Spadix reduced or aborted	Caused by a deficiency of calcium induced by high soil acidity. As root system expands, symptoms no longer appear.
Chilling injury	Flower discolouration: Purple spots on blooms	Control storage temperature.
Spathe blueing	Discolouration of spathe	Exposure to ethylene should be avoided.

3

Bougainvilleas

With its spectacular mass of brightly coloured bracts, the Bougainvillea is unrivalled both in beauty and utility, particularly in the gardens of tropics and subtropics. It is a climbing shrub and in Rio de Janeiro, it also grows as a tree. The Bougainvillea is commonly grown in the open in the gardens in tropical Asia. It is also popular in Africa, Europe and tropical U.S.A. The cultivars of *Bougainvillea spectabilis* and *Bougainvillea glabra* are more tolerant to cooler climate. The Bougainvillea is found growing both as a wild and as a cultivated garden plant in its native home in South America, Brazil, Colombia, Ecuador and Peru. In India, the Bougainvillea grows best in Bangalore, Mysore, Pune, Nagpur, Jabalpur, Gwalior, Chinnai, Hyderabad, Calcutta, Lucknow, Kanpur, Allahabad, Aligarh, Agra, Chandigarh, Patiala, Jaipur, Udaipur and Delhi. In northern India, it flowers during September to December and again during February to June while in south India, it flowers in February-March and August-September. The Bougainvillea generally fails to flower in shade but if it does the flowering is poor and the colour of the bracts is never bright.

The colour is much influenced by the climate. Though the Bougainvillea thrives well in the plains of India, in the hills, a few cultivars like *Bougainvillea glabra* 'Sanderiana' grow well at higher altitudes (from 610 to 1525 metres) above the sea level and even upto 2,285 metres in some places in northern hills like Nainital, Shimla, Almora and also in Nilgiris. The Bougainvillea belongs to the family Nyctaginaceae. *Bougainvillea spectabilis, Bougainvillea glabra* and *Bougainvillea peruviana* are of horticultural importance.

Cultivars

There are numerous cultivars of the Bougainvillea. These have arisen as a result of bud sports or as a seedling variation as a result of chance crossing in nature.

The important cultivar belonging to different species grown in India are described below:

Bougainvillea peruviana

- ☆ **Princes Margaret Rose:** Bracts 2.5 cm × 1.9 cm, claret-rose to fuchsine-pink, flower tube greenish purple. Leaves large, 7.5 cm × 3.5 cm.
- ☆ **Ecuador Pink:** Bract pale magenta pink, crinkle. Flowering best in dry season.
- ☆ **Mary palmer:** A complex chimera with white and magenta coloured bracts on the same plant.
- ☆ **Dr. B.P.Pal:** Bracts large, white. Flowering profuse, Vigorous, suitable for growing in pots.
- ☆ **Shubhra:** Bracts white, floriferous, vigorous.
- ☆ **Fantasy:** A bud sport from 'Princess Margarret Rose'. Emlarging bract greenish with purplish margin changing into tyrian purple or rosein purple with occasional red steaks.

Bougainvillea glabra

- ☆ **Sanderiana:** Plants semi-erect and vigorous, flowering compact and heavy. Leaves 9.6 cm × 4.8 cm, dark green, glabrous and elliptic. Flowering all along the branches. Bracts 3.4 cm × 2.4 cm, cyclamen purple, ovate with acute tip and cordate base. Stars prominent. Suitable for growing in pot, tube or as a hedge. Easy to propagate. Withstands sever pruning can withstand slight frost.
- ☆ **Cypheri:** Plant semi erect, moderately vigorous, compact with short and thin branches. Leaves elliptic, upper surface dull green, 9.6 cm × 7.0 cm. Free flowering flushes. Bracts larger, apex blunt, orchid-purple to petunia-purple, slightly redder when old. Stars prominent useful for growing in pots.
- ☆ **Formosa:** Leaves large, 18 cm × 8.5 cm. upper surface dull green. Bracts 3.9 cm × 2.5 cm. elliptic with a short abrupt-tip, pale rosy manure, chaning into a redder tint when old. Persistent, losing colour on ageing, old bracts

unsightly, flower tube much swollen and strongly angled. Flowering takes place almost throughout the year and as it is very dense on new growth, it is better to prune shoots after flowering.

☆ **Magnifica:** Leaves large, 10 cm × 4.4 cm. elliptic upper surface dark green and shining, sparsely pubescent. Flowering continuous. Bracts borne more closely giving a good mass effect, cyclamen-purple changing into rhodamine-purple on ageing 4 cm × 3.3 cm. Flower tube slightly angled. Suitable for pergolas.

☆ **Mong Heng:** Bracts long, cyclamen purple but slightly and narrower, not persistent, do not change colour on ageing. Grows better in pots.

☆ **Splendens:** Plants and leaves hairy, growth vigorous. Flowering all along the branches. Bracts magenta rose, large, ovate. In Delhi, it flowers in summer only.

☆ **Snow White:** Plants semi erect, quite-quite vigorous, leaves elliptic, dark green, glabrous. Flowering along the branches. Flowering shy in retentive soil and better in light sandy soil. Bracts elliptic apex acute, greenish white when young, changing in the white at flowering.

Bougainvillea spectabilis

☆ **Speciosa:** Vigorous plants, evergreen. Leaves large, 12 cm × 7.7 cm, ovate to cordate, highly pubescent. Bracts 5 cm × 3.5 cm ovate, hairy, apex slightly rounded, magenta rose or orchid-purple, changing little with age, main veins green. Flower tube slightly angled. Free-flowering, suitable for weeding standards.

☆ **Thomasii:** Plant not so vigorous but leaves similar as speciosa. Bracts solferino-purple when young changing into carmine when old, ovate, apex slightly rounded 4.8 cm × 3.8 cm. Free flowering.

☆ **Lateritia**: Plant poor in growth. Leaves 7.5 cm × 5.0 cm., hairy, broadly ovate. Bracts 3.8 cm × 2.6 cm poppy red or red purple when young, brick-red at flowering, fading into orange, elliptic to ovate. Flower tubes hairy. Stars prominent. Flowering best dry and drought conditions.

☆ **Refulgens:** A beautiful climber with drooping branches and long pendulous inflorescences. Leaves 10 cm × 6.6 cm dark green-thick, elliptic, upper surface rough and densely, lower surface smooth. Bracts 4.2 cm × 2.6 cm cyclamen purple or deep purple mauve ovate with cordate base and acute tip, reflexed with prominent incision. Flower tube twisted at the apical end, constricted in the middle, slightly pubescent. Stars prominent, very floriferous. Flowers best in prolonged cool, dry season, suitable for growing in pot and as a standard.

☆ **Maude Chettleburgh:** Plants vigorous. Leaves 10.5 cm × 6.5 cm ovate, with rounded base and short acuminate tip upper surface dark green, pubsely pubescent. Bracts almost violet. Flowers profusely in cool, dry season.

☆ **Cannellii:** Leaves large, 10 cm × 6.5 hairy. Bracts large, 6.2 cm × 4.3 cm, oblong, apex truncate with a short tip, brilliate rose pink. Flower tube hairy.

Uses

1. As climbers
2. As shrubs
3. As bush specimens
4. As hedge
5. On Arches or pergalas
6. As an Espalier
7. Up a tree or on a stump
8. As a standard
9. On slopes and mounds

Cultivars for different purposes

Climbing: Mary palmer, Nawab Ali Yavur Jung, Thimma, Mrs. H.C.Buck, Lady Mary Baring, Isabel Greensmith.

Shrub: Summer Time, Tomato Red, Thimma, Spring Festival, Dr. R.R.Pal, Flame.

Bush: Dr. R.R.Pal, Thimma, Mary Palmer, Flame, Blondie.

Hedge: Partha, Mary Palmer, Thimma.

Arch: Magnifica, Dr. R.R.Pal, Pantha.

Espalier: Lady Mary Baring, Partha.

Standard: Formosa, Natalii, Thimma, Mrs. H.C.Buck.

On slopes: Dr. R.R.Pal, Thimma.

Pot: Flame, Mahara, Mary Palmer, Sonnet, Summer time, Sweetheart, Tomato Red, Lilacina.

Propagation

The Bougainvillea is propagated by cuttings, layerings and budding. Most of the cultivars can be easily propagated by cuttings. Usually, hardwood cuttings, 10-15 cm long and as a thick as a pencil are inserted in the soil in a pot against the rim. The soil mixture in the pot consists of 1 part soil and 2 parts sand. The cuttings strike root in 4-6 weeks. A few cultivars like Mary Palmer, Rosa Catalina, Lateritia, Tomato Red, Louise Wathen and Cypheri are difficult to propagate by cuttings. In some cases like cypheri greenwood cuttings are better than hardwood cuttings. Hormons like IAA, IBA or NAA can be used for better root formation. The cuttings can be set out in July-August or February-March. Cultivars difficult to be propagated by cuttings are layered. Cultivars difficult to be propagated by cuttings or layering can be propagated by T budding. Dr. R.R.Pal in which the cuttings root very easily can be used as rootstock. Budding is done during February-April.

Cultivation

Soil

The plants can grow in most well drained soil and even in small pockets of soil in between rocks. They grow luxuriantly in the light sandy soils of Rajasthan and the red soils of Andhra Pradesh. Once the plant gets established, it sends out its roots deep into the ground, becomes self supporting and requires very little care and attention.

Situation

A sunny site is essential for best growth and flowering. The plants either do not flower or produce only a few flowers when grown in shade. At the time of planting in a garden, careful consideration must be given to the choice of a suitable site. The bright deep-magenta or purple of the more common Bougainvillea cultivars dominates over almost every colour of the flowers growing nearby. It is never advisable to grow roses or annual flowers in front of such cultivars. The Bougainvillea produces best effects when grown against a light coloured or white wall or as a specimen bush in the foreground of a hedge or in a corner of a lush green lawn. The plants thrive well even in hot summer. The young plants are often susceptible to frost and low temperatures during winter, particularly in north western plains and in the hills, the newly propagated plants need protection from frost during winter.

Planting

The best time of planting in during July to September. The normal distance of planting is 2-2.5 metres but when planted for making a hedge, a closer spacing may be given. The pits are dug about 90 cm in diameter and 75cm deep. About 3-4 baskets of well rotton cowdung manure may be added to the soil in each pit at the time of planting.

Nutrition

At the time of planting, a young plant in the ground, 10-15 kg of cowdung manure may be mixed with the soil in a pit of 60-75 cm. Well established plants need little manuring. An application of bonemeal to the soil at the rate of about 250 grams per plant once a year is useful in producing good growth and flowering. Fertilizers can be used for top dressing when the growth is not satisfactory. A mixture of Ammonium sulphate, Super phosphate and Potassium sulphate (1:3:2) applied annually at the rate of 250 grams to a fully grown plan has been found to promote profuse flowering. Organic manures or fertilizers can be given to plants after pruning in June in New Delhi and around. A heavy dose of nitrogen should be avoided as it has an adverse effect on flowering. If the plants are supplied with heavy doses of nitrogen or manure, they tend to produce more vegetative growth, resulting in poor flowering. However, in Florida a high nitrogen dose of 335 kg N/ha annually is reported to promote profuse flowering. Plants respond favourably to liquid cowdung manure applied at fornightly intervals during the initial stages of flowering. In Florida, a complete fertilizer mixture having 6 per cent each of N, P and K and 3 per cent magnesium oxide is applied at the rate of 113 grams per plant several times a year.

Irrigation

The plants do not like irrigating too frequently. The young plants, however, do require regular and frequent irrigation particularly during summer. A heavy and frequent irrigation during flowering often causes shedding of bracts. Cultivars differ markedly in their water requirements.

Pruning

The plants do not require much pruning for best effects. The extent of pruning depends upon the cultivar and its use. It is necessary to prune the shoots and stems when the plants are grown as a hedge, arch or a pergola or a hand rail. Trimming of the sides and the top is common in a hedge. Pruning is also done on plants grown as standards, espaliers or grown in pots. In standard, it is essential to remove all the side branches on the main stem and keep the umbrella like top in good shape. Side branches and extra growth are pruned when the plants are trained on an espalier system. Specimen bushes grown on a lawn and those growing on banks or on tall trees do not require any pruning. In New Delhi, Bougainvilleas should be pruned in May or June after the end of the flowering period. Pruning should not be done after rainy season as, it would discourage flowering. Tipping should be completed before the middle of September. A light application of manures and fertilizers after pruning is beneficial to the plants.

Pests and Diseases

A caterpillar, *Asciodes gordinalis* and a beetle, *Amphicerus cornatus* are common in Florida which feed on leaves, flowers and burrows into plants. Leaf spot (*Cercosporidium bougainvillea*) spraying of plants with some copper fungicides @ 0.2 per cent at 7-10 days intervals. *Cladosporium arthrinoides* leaf spot and an unknown virus have been reported from Florida. Soft scale (*Coccus hesperidum*) can be controlled by spraying of Malathion. Sometimes chlorosis or yellowing of leaves is observed in pots and can be corrected by foliar application of urea. Chlorosis due to Mn, Cu, Zn, Fe deficiency may also occur.

4

Calla Lily

Zantedeschia commonly known as calla lily is a perennial herb grown for their ornamental corolla like spathe and sometimes for their variegated or spotted foliage. It was introduced to Europe in the seventeenth century as *Zantedeschia aethiopica* and is now widely naturalized including Europe, North America, Central America, South America, Oceania and Australia. *Zantedeschia* is represented in North America primarily as cultivars used as ornamental house plants. Eight species are currently recognized: *Zantedeschia aethiopica* 'arum lily', *Zantedeschia albomaculata* 'spotted lily', *Zantedeschia elliottiana* 'golden calla', *Zantedeschia macrocarpa* 'trumpet-shaped', *Zantedeschia oculata* 'yellow calla', *Zantedeschia melanoleuca* 'black-throated calla', *Zantedeschia rehmannii* 'pink or rose calla' *and Zantedeschia valida.*

It has often been used in paintings and is visible in many of Diego Rivera's works of art. Zantedeschia or Calla lily is a very beautiful flower. Flowers delivered the feelings subtly and every part of gifting a flower, carried secret flower meanings.

Calla lily was used to express many such hidden symbols. These can be planted along ponds, lakes or in bog gardens. The colourful flowers and leaves are highly valued. Species and cultivars are widely used as ornamental plants and for cut flowers. Callas are naturally a warm-temperate climate crop, however, are grown in many locations around the world. Cooler climates require heat retaining greenhouse structures to achieve satisfactory growth and economic time to flowering. Callas are also very successfully grown in equatorial conditions at high altitude of 1800-2400m in warm sunny days up to 25-30°C and nights of 6-10°C. Optimum crop performance is dependent on temperature in the crop canopy and rhizome depth. Ideal day temperatures are 18-28°C and night 12-18°C. Optimum soil/media temperature is 18-20°C at rhizome depth (5-10cm) with an upper limit of 23°C. Light is a critical factor in calla production. Over shading or low light may result in low flower yields, dull colours, poor flower performance, plant stress resulting in soft rot and significantly lower rhizome multiplication. Callas perform better in rising light conditions and longer day length.

Distribution and Habitat

All species are endemic to central and southern Africa, from Nigeria to Tanzania and South Africa. *Zantedeschia aethiopica* grows naturally in marshy areas and is only deciduous when water becomes scarce. It grows continuously when watered and fed regularly and can survive periods of minor frosts.

Soil/Growing Media

Free draining fertile soil has been the most common growing medium but there are significant limitations for continuous cropping. Check previous use including crops grown, pre-emergent chemicals and fertilizer used and always test soil for nutrients and pathogen status prior to planting. Growing media may include some or all of the following: washed coir fibre, peat, composted pine bark (fines and fibre), perlite/vermiculite and rice husk. Media is normally renewed or sterilized after each growing cycle. Another option is to grow rhizomes in pine sawdust or rice husk either over a layer of soil/media (separated by net) or on its own in a hydroponic system.

Greenhouse Production

Simple well ventilated plastic rain cover type greenhouses (plastic tunnel) are best for controlled calla production and essential for cut flower cropping in wet conditions and tissue culture plant establishment. More expensive heat retaining structures are required for cooler climates and autumn/winter production. Cover greatly improves crop quality, timing and reduces rhizome losses. Ability to regulate soil moisture gives greater control over rhizome maturation and rhizome harvest timing. In dull or winter conditions high light transmission is essential keep greenhouse film or glass clean. Ventilation and air movement is critical. Circulating or horizontal fans are recommended. Use of moveable shade is beneficial in hotter conditions. Thirty + fifty per cent shade after the early vegetative stage; during flowering; and hot days during the latter rhizomeization stage can significantly reduce stress and possible infection. A mulch layer of clean pine sawdust or rice husk/straw is used to reduce soil heat, retain moisture and reduce weeds.

Propagation

Conventionally multiplied through rhizomes. Quality plant material derived from verifiable genetic stock is essential. First year T1 rhizome explant tissue culture (TC) or second cycle T2 stock are the most vigorous, disease free and reliable starting material. Older larger rhizomes may not acclimatize well in a new enviro-location and incur high losses. These rhizomes will last 4-5 growing cycles when managed well. A 25 per cent annual replacement of TC or T1 rhizomes is essential for an ongoing commercial cut flower project. Many larger growers start directly with TC.

Selection of Cultivars and Colours

The market (auction, wholesaler etc) and grower experience will help to decide the best varieties and colour balance - white, yellow, golden, orange, peach, pink, red, lavender, purple and black. Choose cultivars for their productivity, colour and end use - cut flowers, pot plants or landscaping. Several cultivars are: Little Gem (plant 30-40 cm high, spathe white, 8-12 cm long), Devonensis (dwarf, free bloomer, fragrant), Candidissima (spathe large, pure white), Gigantea (very large plant), Godfreyana (dwarf, white spathe), Grandiflora (large spathe) and Chlidsiana (dwarf and free flowering).

Bed Preparation

Cultivate soil 2-3 months prior to planting. Ensure that the soil has a good free-draining structure, with adequate air fill porosity. Form raised beds (200mm) that fit tractor wheel spacing and other machinery. Standard bed width is 1-1.2m with 40-50cm walkways.

Sprouting of Rhizomes

Sprouting promotes even crop emergence, better Gibberellic Acid (GA) absorption and shortens the growing cycle. This process is vital if rhizomes have a short dormancy (10-12 weeks after lift), and when growing in cooler conditions. Prior to planting and application of GA, remove rhizomes from storage and pre-germinate at 21-25°C and 85 per cent RH for 7-10 days to trigger emergence of growing shoots. Smaller 2 cm rhizomes have much less ability to withstand temperature and humidity fluctuation - these are normally not pre-germinated. Once the growing shoots have emerged a minimum of 5-10mm, apply GA prior to planting. This will significantly increase flower initiation when applied correctly. GA_3 is distributed as tablets or powder and is applied by spraying to runoff or dipping for 15 minutes at 100-125 ppm only when shoots have emerged at least 1cm. GA is much less effective on dormant rhizomes with no shoots.

Spacing

Rhizome spacing depends on growing conditions, availability of space and grower preference. Lower densities often result in more flowers per rhizome, larger flower heads and greater rhizome multiplication.

Planting

Planting of crop is dependent on when flowers and pot plants are to be marketed. A cut flower programme of continuous 2-3 weekly plantings keeps product in the market and promotes the grower. Planting in late winter/early spring with early greenhouse forcing or a late crop at the end of summer can be very profitable, however, for best flower production and rhizome multiplication, spring planting is the best option. Avoid planting in severe heat. Irrigation just after planting allows settling rhizomes into the media. Apply sawdust/rice husk mulch before shoot emergence.

Manures and Fertilizers

Addition of composted organic material can assist nutrient holding capacity and rhizome growth. A base dressing of N, P, K, Mg plus trace elements will supply the necessary balance of nutrients for plant growth. Foliar feeding should only be used as a remedial backup as is much less effective than liquid feeding. Avoid high Nitrogen (N), especially later in the crop cycle as it may promote lush vegetative growth prone to plant stress. Liquid fertilizers like N @ 125-150ppm, P @ 30ppm, K @ 125-175ppm, Ca @ 60-70ppm and Mg @ 20-25ppm. At the end of flowering keep feeding for at least six to eight weeks progressively adjusting N-K ratio to 1:3 to maximize rhizome growth. Nutrition during flowering like Fe may have an effect on colour intensity.

Irrigation

Clean water with an added residual bacteriostat is a proven component of achieving optimum calla production and controlling plant mortality/soft rot. Actual water volume depends on soil/media drainage capacity and evapo-transpiration. Reduce irrigation before the harvesting of rhizomes. Irrigate in the morning, prior to the heat of the day, mist to reduce heat and irrigate again as required in late afternoon. Ground level drippers conserve water and provide more uniform watering in windy conditions.

Weed Control

Weeds affecting soil grown crops are best eradicated 3 months prior to planting when in active growth. Spray Roundup or Glyphosate (10 ml/litre) and leave for at least 7-10 days before cultivation. Proven pre-emergent herbicides include **Simazine** and **Alachlor.** Apply after planting time prior to rain/irrigation. Never spray in the heat of the day as burning may occur and ensure soil moisture is sufficient to activate the chemical ingredient after application.

Harvesting and Pre-cooling of Flowers

Harvesting of flowers in the cool morning. In dry conditions irrigate prior to harvest. Flowers are pulled rather than cut to ensure the longest possible stem length. Place harvested stems, dry in cool store at 6-8°C for pre-cooling prior to grading. Stems can stay dry for up to 6-8 hours without any damage.

Post-harvest Handling

After pre-cooling, grade and bunch (5 or 10 stems) per market requirements. Cut stems with a clean sharp blade in the basal white part of stem to help to avoid splitting and roll-up of the stem. Prepare a hydration solution of clean water or chlorine to protect against slimy stem *(Erwinia)* and increase vase life. Do not use sugar based preservatives. Re-cut and place in fresh solution for longer storage. Cool store flowers at 6-8°C minimum 80 per cent RH prior to packing.

Packing of Flowers

Remove bunches from hydration solution and remove excess moisture with clean paper tissue to help to avoid slimy stem or botrytis in transit. Final packing into cartons should be as per market preference. Do not over pack flowers as damage may occur. Secure bunches in carton with rubber bands or tape to avoid movement in transit.

Post-harvest Practices of Rhizomes

Rhizome Maturity and Lifting

Commence lifting when leaves turn yellow and die down, roots regress and the rhizome surface becomes tougher. Harvesting any earlier will reduce rhizome multiplication and maturation. Immature rhizomes are easily bruised and can suffer breakdown in storage or calcification (death of rhizome tissue) within several weeks. Lift rhizomes carefully either by hand or mechanical digger. Leave roots on during initial drying to seal up rhizome and protect from invasive pathogens.

Washing and Dipping

Rhizomes should be lifted, cured and dried with minimal handling. If soil/ media consistency allows do not wash or dip rhizomes. *Mucor* (fungal rot) can attack rhizomes in storage. It can arise after lifting during wet conditions and/or high humidity in early storage. Control of this fungal disease with **Ridomil** and **Fungiflor**.

Rhizome Curing

Although rhizomes can be lifted and cured/dried in a greenhouse, this is difficult where large quantities of rhizomes are harvested on a daily basis. Cure rhizomes for 2-5 days at 20-25°C immediately after lifting, ensuring good air movement is maintained at all times.

Rhizome Dormancy and Storage

Callas require at least 10-12 weeks rhizome dormancy after lifting. Longer term cool storage is preferable enabling faster germination, consistent shoot emergence and better flower production. Optimum RH: 65-70 per cent.

Grading of Rhizomes

After 3-5 weeks of stabilization cured rhizomes can be de-rooted, graded and sorted ready for storage to complete dormancy. After handling rhizomes, re-cure overnight prior to further storage to seal any newly damaged or exposed tissue.

Storage

Rhizomes can be stored for 6-10 months at 8-10°C and 70-75 per cent RH. Prior to loading the store, clean thoroughly with sanitizer and check all cooling and air circulation equipment.

Plant Protection

Insects

The main pests are thrips and aphids. Control is especially important to halt the spread of virus and to maintain flower quality for export phyto-sanitary purposes. Mealy bug (*Pseudococcus* spp.) can attack the growing crop in the latter stages and leave eggs that hatch post storage, as the rhizome is warmed up. Insecticides *viz.*, Diazinon and Imidacloprid should be applied from spike emergence and repeated at 7-10 day intervals up to flowering.

Fungal Control of Rhizomes

Callas can be attacked by a range of primary fungal pathogens including *Pythium, Fusarium, Rhizoctonia* and *Phytophthora*. These often attack the root zone and may not exhibit above ground symptoms until well established up to 7-10 days later after infection. Withering, rolled up leaves are often a symptom of below ground problems. A drench of **Ridomil** (1 ml/10m^2) alternating with **Thiram** (3ml/l) is effective. **Captan** (1.5g/l) may be used as an alternate to **Ridomil**, effective against pythium and phytophthora.

Fungal Control of Flowers

Fungal spotting on flower-heads during humid weather or rain is common. The main fungal diseases are *Botrytis, Acremonium* and *Alternaria*. Preventive sprays should be applied every 7-10 days from bud emergence: **Captan, Thiram** and **Mancozeb** at label rates. Each individual chemical should only be used 2 - 3 times per season.

5

The Canna

Canna is one of the most versatile ornamental plants valued both for its foliage and flowers and is popularly known as "Indian Shot". In earlier days canna was almost entirely grown for its attractive foliage and deep green or deep coppery, red or green edged with bronze colour. In England and Europe, people used to grow canna to give a tropical look to the landscape with lush foliage. Canna belongs to the family Cannaceae. The green canna is a native of tropical America and Asia and comprises over 50 species. Among them, *Canna indica* Linn., *Canna coccinea, Canna speciosa* Roscol, *Canna nepalensis, Canna orientalis, Canna flaccida, Canna reevesii* Lindi, *Canna glauca* Linn., *Canna edulis, Canna werscewiczii* and *Canna iridiflora* Ruiz and Pav. are most important.

Cultivars

Canna has got a number of cultivar, with most of them having variations of yellow, red, pink colours. Some cultivars have simple colour, whereas, others have splashes, margins, sopeckles and centres of other colours. Some of the important cultivars are:

1. Abhiskar – Glowing shade of rosy-scarlet.
2. American Beauty – Orange scarlet.
3. Apricot – Buff-yellow base overspread with salmon pink suitable for bedding.
4. Assant – Scarlet flowers and dark bronze foliage.
5. Black knight – Crimson-maroon flowers with dark bronze folige.
6. City of Portland – Large flowers with deep glowing pink colour.
7. Damador – Bright rich ruby-pink self colour flowers.
8. Edith – Intense orange flowers.
9. Golden Wedding – Yellow flowers.
10. Golden Dawn – Golden yellow flowers.
11. King Humbert – Glowing orange-scarlet with red markings-leaves are red bronze in colour.
12. Lady Bird – Soft yellow spotted with small red dots.
13. Percy Lancaster – Yellow with heavily spotted red.
14. Statue of Liberty – Blazing flame-red flowers. Foliage bronze purple colour.
15. Ambassador – Brilliant cherry-red colour-good for bedding.
16. White Queen – Creamy white flowers.

Propagation

Canna is easily propgated by division of rhizomes *i.e.*, rootstock. It is better to cut the rootstock into such a way that each piece has several strong buds. These pieces may be planted directly in the beds, one bud pieces are suitable for planting in pots as they produce weak plants. Where strong mass effects are desired, it is better to plant entire stool as such.

Planting

In the plains of India, the best time of planting cannas is just before the onset of monsoon. In the hills, planting is done sometimes in March-April. Isolated plants of canna do not give a satisfactory effect. Therefore, it is advisable to put them in small clumps in the hardy border or the shrubbery. Most striking effect comes when they are planted in mass with tall cultivars in the centre and dwarf ones sloping onwards. Plantation of cannas in formal beds gives equally as charming look; normally against a heavy background of green, the bright and strong colours show to their best. While planting care should be taken neither to plant too deep or too shallow. The rhizomes should be atleast 5-7 cm under the soil. Rhizomes which are infested or damaged should be discarded. The plant to plant and row to row distance should be maintained between 40 and 50 cm and the plants in two rows should be at alternate positions rather than in straight lines. After planting, the soil should be processed hard and irrigated. It is advisable to replant the entire bed every year to have better blooms and to avoid overcrowding. For this, the stools should be lifted just ahead of monsoon and the beds prepared as for fresh planting. While the beds being prepared, the stools may be kept under shade. From the time, a new rhizome is planted, it takes about 6 weeks for the flower spike to appear.

Storage of Rhizomes

After the top have dried, just by the end of autumn, the rhizomes are lifted. Alow them to dry and then shake off any surplus soil sticking around retaining some of it on the roots. Store them on the shelves in single layers in a warm, dryroom. Take care that the roots are not damaged while in storage due to overwarming of room or excessive moisture. Yet another indigenous way of storage is to dig a pit in a slightly protected place about 1 metre deep. After proper drying, the rhizomes are placed in the pit, covered with a thin layer of soil and the pit itself is covered with thatch, twas, leaves or straw and covered with the soil.

After Care

Weeding and hoeing operations are carried out regularly. Regular forking of the soil should be done. Spent up stalks, dried leaves and faded flowers should be promptly removed every day. For this purpose a sharp knife or a Secateurs are used. While irrigating, care should be taken than excess water do not accumulate in the beds. The plants should be sprayed with water to wash the dust and to provide flesh look.

Preparation of Beds

Canna gives the best effect when raised in beds rather than in pots. It requires a sunny place and does not flower well in shade. The beds for canna should be worked deep and clods etc, broken. Expose the dug up beds with sunlight for a few days before mixing of manures and their dressing up. Canna is a heavy feeder and as such at least a 5-7 cm thick layer of well rotted F.Y.M. should be worked in. Adding of about 100 g of Superphosfate and 50 g of Potassium Sulphate/sq. m of bed will help in the better growth of plant. The beds should be lightly irrigated and allowed to settle for a week or so before the actual planting is taken up. The soil of the bed should be friable, well drained, rich in humus and clayloam in texture. Where the oils are not well drained, the rhizomes start rotting due to water stagnation.

Pests and Diseases

Rats sometimes become a nuisance when they burrow holes and damage the rhizome. A minor damage is done by caterpillars and grass hoppers which feed on the leaves. Birds often attracted by the bright colours of flowers damage them. Among diseases, clump rot, sometimes causes damage. The trouble is indicated when the older leaves flag and the green or bronze colour of foliage fades with new shoots becoming black. It is avoided by removing the clump and soak in a solution of suitable fungicide for sometimes. The soil of such a bed should be exposed to strong sun and treated with formalin.

6

Carnation

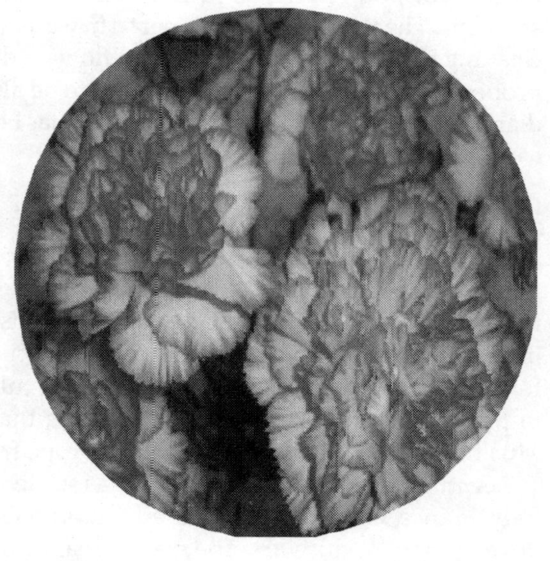

Carnation (*Dianthus caryophyllus* L.) is one among the most popular commercial cut flowers of the world, ranking second in commercial importance next only to rose. Carnation is preferred on account of its excellent keeping quality, wide range of forms and colours and ability to withstand long distance transportation. Cut carnations, roses and chrysanthemums contribute close to 50 per cent of the world cut flower trade. Carnation belongs to the family Caryophyllaceae. The genus name *'Dianthus'* is derived from the Greek words 'dios' meaning 'God' or 'divine' and 'anthos' meaning 'flower' and hence known as 'Divine Flower'. The species name *'caryophyllus'* is derived from the Greek word 'caryan' meaning 'nut' and 'phyllon' meaning 'leaf'. The name *'caryophyllus'* has been chosen by Linnaeus after the genus name of clove due to the clove-like fragrance of carnation. The common name 'carnation' probably must have come from the Greek word 'coronation' because these flowers were used in decorating the crown of Greek athletes. Carnation is the national flower of Spain. Carnations are excellent for cut flowers, bedding, pots, borders, edging, indoors and rock gardens. They give a unique softness to the rock gardens. Though cut carnations are traded in the world market year round, they are in particular demand for the Valentine's Day, Easter, Mother's Day and Christmas. Miniature carnations are now gaining popularity for their potential use in floral arrangement. They were known as 'Jove's flower' in ancient Rome as a tribute to one of their beloved gods. Carnations are associated with some sentiments and symbolisms. In Korea red and pink carnations are used in expressing love and gratitude towards parents on Parents' Day. Pink carnations are symbol of mother's undying love. Carnations are also used on Teachers' Day to express admiration and gratitude to teachers. Light red carnations represent admiration, while dark red ones denote deep love and affection. White carnations represent pure love and good luck; striped carnations symbolize regret that a love cannot be shared. Green carnations are used as a secret gay code, while purple carnations indicate capriciousness.

The flower petals of carnation are candied used as a garnish in salads and for flavouring fruit, fruit salads. They are also used as a substitute for rose petals in making syrup after removing the bitter white base. Carnations are commercially utilized for extraction of perfume in France and the Netherlands. The volatile oil of carnation contains 40 per cent benzyl benzoate, 30 per cent eugenol, 7 per cent phenylethyl alcohol, 5 per cent benzyl salicylate and 1 per cent methyl salicylate. About 100 g of oil is obtained from 500 Kg of flowers. The flower heads are dried and used in pot-pourri, scented sachets and cosmetic products. The plant is quite rich in saponins. The leaves can be simmered in water and this water can be used as soap for cleaning skin and clothes. Carnation flowers due to their aromatic and stimulative properties. The flowers are considered to be alexiteric, antispasmodic, cardiotonic, diaphoretic and nervine. The whole plant has been used as a vermifuge in China and as an animal feed in Spain.

Origin and Distribution

The *Dianthus* species are adapted to the cooler alpine regions of Europe and Asia and are also found in the Mediterranean coastal regions. This is consistent with records on floras indicating that the natural distribution of carnation is restricted to the Mediterranean regions of Greece, Italy, Sicily and Sardinia. It has been cultivated for over 2000 years and today commercial cultivation is the result of 200 years of improvement and breeding. It is believed that carnations were cultivated by the Muslims of Africa and introduced to Europe from Tunis in the 13th century. During 16th century, its improvement work was started in various countries. The gardens of Italy, France, Germany, the Netherlands and England have contributed much to the development of cultivars. The major breakthrough in the carnation flower industry came with the evolution of the cultivar William Sim in 1938 by William Sim of USA. From this red variety there have been mutations to white, pink, orange and different variegated forms. The first genetically modified carnation variety 'Moon Dust' was developed in Australia by a plant biotechnology company called 'Florigene'.

In India, Sim carnations are reported to have been introduced first by the Maharaja of Patiala at his farm at Dochi. The natural climates for carnation cultivation occur near 30°N or S latitudes of the equator. Carnation growing countries are Spain, Kenya, Columbia, Israel, Ceylon, Poland, the Netherlands, France, Germany, Italy, Canary

Islands, Australia, Valparaiso, Chile, USA and South Africa. An altitude of 2000-2500m is ideal for carnation cultivation. In India, potentiality for growing good quality of carnation remains in cool climatic areas like Kashmir, Kullu Valley, Kalimpong and Bangalore. The area around Delhi, U.P. and Punjab, Nasik, Srinagar, Solan, Coimbatore including Nilgiri are the promising carnation growing zones in India. Carnation cultivation is also increasing in the area around Bangalore and Pune.

Botany

Carnation is a semi hardy herbaceous perennial with thick, narrow, linear and succulent leaves. Leaf blades are simple, entire, linear, arranged in pairs, their colour varies from green to grey-blue or purple. The stems are hardy, shiny and have one to three angles with tumid joints. Each stem produces a terminal flower and hence inflorescence is generally a terminal cyme, sometimes racemiform. Flowers are bisexual and occasionally unisexual. The flower colour varies from white to pink or purple in colour. Normally flowers between 6 and 8.5 cm in diameter. Some disbudded greenhouse grown plants for exhibition have flowers up to 10 cm diameter. Petals are broad with frilled margins and calyx is cylindrical with bracts at the base. The stamens can occur in one or two whorls, in equal number or twice the number of the petals. The fruit is in the form of a capsule and contains many small seeds. The fruit ripens within five weeks of pollination. The fruits contain an average of 40 seeds. On maturity, the tubular capsule opens from top and releases the seeds. The basic chromosome number in Dianthus is 15. Carnations are generally diploids (2n=30), though tetraploid forms (4n= 60) have also been identified. Triploid carnations were produced for commercial purpose but the resulting plants were mostly aneuploid. The majority of cultivable carnations are diploid. Flower colour in carnation is attributed to the presence of two major pigments *viz.*, carotenoids and flavonoids. The carotenoids are responsible for colour ranging from yellow to orange. Anthocyanins which contribute colour to carnation flowers red or magenta colour and the pelargonidins which are responsible for orange, pink or brick red colour.

Species

The family Caryophyllaceae consists of 80 genera and 2000 species which are either annual or perennial and most of them occur in the northern hemisphere. The genus *Dianthus* has about 300 species of which only a few are cultivated *viz., Dianthus caryophyllus, Dianthus barbatus* and *Dianthus chinensis.* Modern day perpetual flowering carnation is a cross between *Dianthus caryophyllus* and *Dianthus chinensis.* The species *barbatus* is commonly known as 'Sweet Williams', which grows readily from seeds. The species *chinensis* is commonly known as 'Indian Pink' or 'Japanese Pink'. This species is excellent for beds, borders, edging, rock gardens, pots and cut flowers. The two most commonly grown varieties of this species are the Japanese Pink (*Dianthus chinensis* var. *heddewigii)* and the Fringed Pink (*Dianthus chinensis* var. *lacinatus).*

Types of Carnation

Based on the availability of large number of varieties and diversified cultural requirements, carnations are classified as Chabaud or Marguerite, Border and Picotee, Malmaison and Perpetuals.

Chabaud or Marguerite

These are annual carnations developed by crossing of *Dianthus chinensis* with *Dianthus caryophyllus*. Flowers are single or double, propagated by seeds. Flowers are large with fringed petals and have shorter post harvest life. The various kinds of Chabaud are Giant Chabaud, Compact Dwarf Chabaud, Entant de Nice, Fleur de Camelia and Margarita.

Border and Picotee

The flowers of border type carnations are symmetrical and are the easiest to grow. Border carnations are further subdivided according to colour of flowers as Selfs, Flakes, Bizarres and Fancies as indicated below:

- ☆ Selfs: The flowers are of a single colour
- ☆ Flakes: The flowers have a ground colour striped with another shade
- ☆ Bizarres: The flowers have a ground colour marked and flaked with two or three other tints
- ☆ Fancies: The flowers which do not come into the above subdivisions

Malmaison

These are strong, sturdy and stiff plants with broad leaves. Flowers are large, double with well filled centres and are mainly pink coloured with good fragrance.

Perpetuals

These are hybrids of different *Dianthus* species. The plants are not hardy and are generally treated as cool greenhouse plants. They bear flowers round the year with long stalks which make them suitable for cut flowers. They produce better quality flowers and withstand long distance transportation. These perpetuals are classified into two classes *viz.*, standard and spray carnations.

Standard Carnations

Standard carnations produce larger blooms with longer stems, usually a single large flower on an individual stem. Carnation cultivars differ in the length of the vase life of cut flowers which is one of the characteristics determining the commercial value of the ornamental flowers.

Standard Cultivars

Table 6.1: Flower Colours in Standard Cultivars of Carnation

Flower Colour	*Cultivars*
Red	Guapo, Leopardii, Domingo, Master,Gaudina, Taureg, Turbo.
White	Baltico, Madame Collette, Emotion, Angelica, White Dona.
Light pink	Charmant, Cipro, Big Mama, Tonic,Golem, Dona.
Yellow	Kiro, Salamanca, Soto, Diana, Hermis.

Contd...

Table 6.1–*Contd...*

Flower Colour	*Cultivars*
Orange	Malaga, Florgore, Solar, Star, Orange Prestige.
Bicolour	Spencer, Swing, Madras, Giampi, Naxos, Guinea, Leila, Olympia, Wizard, Navidad, Happy Golem.
IIHRP - 1	Mutant developed using EMS at 0.25 per cent concentration. Red flowers with smooth edged petals. 65cm length. Good keeping quality of 10-12 days. It is tolerant to nematode and fusarium wilt. Yields 300-360 flowers/m2/year

Spray Carnations

Spray or miniature carnations produce smaller sized blooms with shorter stems in bunches. The flowers are borne on short branches of a single stalk.

Spray Cultivars

Table 6.2: Flower Colours in Spray Cultivars of Carnation

Flower Colour	*Cultivars*
Red	Red Eye, Red Fuego, Red Vital, Aveiro
White	White Prestige, Milky Way, Elvis, T-587
Pink	Rosa Bebe, Spur, Suprema, D- 925, Celebration, Osiris
Yellow	Stella, Prestige, Mila, Sonia, Abril
Orange	Sunshine, Autumn, Fancy Fuego, Disney, Eilat
Bicolour	Berry, Orbit Plus, Nadeja, Picaro

Climate and Soil

Climate

Carnations grown outdoors as annuals in beds, border and pots are usually raised from seeds. When the climate is mild and almost free from rains, the seedling plants flower in 120-130 days after germination. For commercial cut flower production, carnations are grown in greenhouses by maintaining optimum growing environment. The growth and flowering are usually influenced by plant genotype, light, temperature, carbon dioxide concentration of the environment and cultural practices. The optimum night temperature for carnation is 10-11°C during winter and 13-15.5°C in summer. The optimum day temperature range is 18-24°C. High day and night temperatures, especially during flowering of carnation result in abnormal flower opening and calyx splitting, very low temperature delays flower bud development markedly. Carnation is a quantitative long day plant. Long photoperiod usually promotes flowering while short days tend to delay it. Flower quality can be improved by providing long days only for a short period (4-6 weeks) when the shoots have 4-7 pairs of leaves. The critical photoperiod for most standard and spray carnations is about 13 hours. For commercial cultivation, the humidity of greenhouse should be maintained at 80-85 per cent during beginning of vegetative growth and 60-65 per cent during full growth stage. Free circulation of air inside the greenhouse is essential. In hills, natural vents have to be provided on the sides or roofs, whereas, in plains, a fan-pad cooling system will cater to the needs of air circulation. A ventilation of 25-30 per cent of the polyhouse ground area is ideal. The best quality flowers can be produced when CO_2 concentration in the greenhouse is maintained at 500 - 750 ppm.

Soil

Sandy loam soils rich in organic matter content with pH of 5.5-6.5 are most ideal for carnation cultivation. Clay and silt soil can be improved by incorporating organic matter or compost. A soil EC of 0.8 - 1.2 dSm^{-1} during the vegetative stage and 1.2 - 1.5 dSm^{-1} during the generative stage is most ideal for carnation cultivation. The soil must be well drained because the crop is highly susceptible to fusarium wilt.

Propagation

Carnation can be propagated both by seeds and vegetative means. Seed propagation is normally practiced to raise marguerite type carnation and for the purpose of hybridization. Cut carnations are multiplied vegetatively by means of terminal cuttings. Micropropagation techniques have also been standardized for large scale multiplication of disease free plants. Seeds are used for crop improvement programme and for raising annual types. Seeds are sown during July-August and also extended up to October. In the hills, sowing is done in two seasons, August to October and March to April. The optimum temperature for seed germination is 21°C. Seeds germinate in about 5-8 days and at four leaf stage, the seedlings are transplanted in nursery pots. Commercially, carnation is multiplied through terminal cuttings. Maintaining healthy mother plants in good vegetative condition is essential. Terminal cuttings of about 10-15 cm with 4-5 pairs of leaves are harvested for multiplication. Cuttings taken from healthy plants fertilized with adequate nitrogen have been rooted better. Cuttings can be stored at 1-3°C for several weeks before rooting. Full sunlight with misting promotes root formation in cuttings.

Treatment of cuttings with a combination of fungicides such as Dithane M-45 and Bavistin each @ 0.1 per cent for half an hour before planting reduces the spread of fungal diseases during rooting. Treating the basal end of cuttings with NAA @ 500-1000 ppm before planting improves rooting of cuttings. Sand, perlite, vermiculite, cocopeat, sphagnum moss are used as rooting media for carnation cuttings. Mixtures of two or more materials are also used. In general, 20-35 per cent, moisture content, 60 per cent porosity, 30-40 per cent aeration and a pH ranging between 6.0 and 6.8 are the most favourable physical conditions of the rooting medium. The medium should be sterilized with formalin before planting of cuttings. Protrays filled with growing media such as cocopeat are normally used for rooting of cuttings. Planting time has significant influence on rooting of cuttings. The root formation is always better during cooler months. Cuttings normally develop sufficient root system within 21 days of planting. After rooting, the cuttings should be transferred to a hardening chamber containing sterilized media made of sand: soil: FYM in equal proportions.

Micro-propagation

Large scale multiplication of carnations can be done through tissue culture. The main aim of in *vitro* propagation is production of disease free planting material. Shoot tips are mainly used for micropropagation.

Land Preparation

Soil for growing carnation should be deeply ploughed to not less than 60 to 70 cm depth. Organic matter should be added to improve the aeration and fertility of soil. Generally, well decomposed FYM @ 25 kg, leaf mould @ 25 kg Neem cake @ 500g and bone meal @ 200g per m^2 can be added. Adding cocopeat (3kg), humic acid granules (5g), seaweed granules (5g) and micronutrients (3g) per m^2 can improve the texture and nutrient status of the growing medium. Steam sterilization involves passing of aerated steam at 60-72°C through the soil covered with good quality plastic sheets. Of late, either Dazomet (30 - 40 g/m^2), formaldehyde (2 litre/m^2 of a solution prepared with 1 litre/7 litres of water) or H_2O_2 (300ml/m^2) is commonly used for soil sterilization. The soil is moistened first and then drenched with the fumigant and immediately covered with plastic. After 48 hrs the plastic cover should be removed and the soil should be thoroughly turned for proper aeration. The fumigant should be removed by leaching with plain water 3-4 times with turning of soil each time.

Bed Preparation

Generally, the basal fertilizer dose of single super phosphate @ 200 g/m^2, potassium sulphate @ 150 g/m^2, magnesium sulphate @ 50 g/m^2 and borax @ 2 g/m^2 should be evenly spread and thoroughly mixed with the media before bed preparation. Apart from the above fertilizers, bio-fertilizers and bio-control agents for the control of pests and diseases can be incorporated to soil at the time of bed preparation. Azospirillum, Phosphobcteria, *Trichoderma viridi, Pseudomonas fluorescens,* VAM each 1 kg can be added for 500m^2 area for enriching the soil. Bed layout depends on the orientation of the greenhouse. However, balanced development of the crop occurs when the beds are formed in the North - South direction. If the beds run East - West, the crop tends to crowd in the northern side. The ideal bed width and height are 75 - 100 cm and 30 - 45 cm, respectively. The bed length should not exceed 25 m. A path width of 45 - 50 cm is ideal.

Planting

Planting time can be chosen depending on market demand. However, planting every 3-4 months is advisable to ensure regular supply of flowers. Before planting, drip lines and support netting are to be laid out. The bed should be moderately wet and drenched lightly with a fungicide (copper oxychloride or copper hydroxide @ 2 g/l) before planting. **25 days old rooted cuttings best for planting.** Planting should be done preferably in the evening. A spacing of 15cm × 15cm is followed. This will accommodate 30 - 33 plants/m^2 considering 75 per cent as the net cropping area. In general four row or six row system of planting is adopted. The plants should be removed from the poly bags or rooting trays carefully without damaging the roots. Planting should be done at shallow depth with part of the root zone exposed. Deep planting will lead to rotting.

Grow-bag System of Cultivation

Grow-bag system is a recent innovation in carnation cultivation. Poly grow bags are made up of specially formulated plastic which has a longer life. The bags are available in various dimensions. The bags are filled with growing media, preferably inert media like cocopeat and used for planting. Grow-bags are recently being used in some parts of the Nilgiris for carnation cultivation. The grow-bag system of cultivation offers the advantages of saving of time and labour required for soil preparation. amenable for drip irrigation and fertigation. Helps to avoid repeated use of the same soil in the green house. High water use efficiency resulting from better water conservation. Complete weed control enhanced flower yield and quality.

Table 6.3: Fertigation Schedule for Carnation

Nutrients	*Quantity in g/m²/week*
Till bud formation	
Tank-A (Monday and Thursday)	
Ammonium nitrate	3.0
Potassium nitrate	5.0
Monoammonium phosphate/Monopotassium phosphate	2.0
Magnesium nitrate (11:0:0:16 of N:P:K:Mg)	2.5
Boron	1.0
Trace elements/micronutrients	1.0
Tank-B (Tuesday and Friday)	
Potassium nitrate	5.0
Calcium nitrate	8.0
Bud formation to harvest	
Tank-A (Monday and Thursday)	
19: 19: 19	2.0
Potassium nitrate	7.5
Monoammonium phosphate/Monopotassium phosphate	2.0
Magnesium nitrate (11:0:0:16 of N:P:K:Mg)	2.5
Boron	0.1
Trace elements/micronutrients	1.0
Tank-B (Tuesday and Friday)	
Potassium nitrate	5.0
Calcium nitrate	9.0

Plant Supports

Both standard and spray carnations require support. Lack of support leads to bending of stems ultimately causing decline in market quality. Poorly supported branches easily break, making them susceptible to diseases. Various kinds of plastic mesh, string, bamboo canes are used for plant support. In recent times, nets are used for supporting carnation plants. The nets are usually laid out in 4 or 5 layers. Before planting, the first layer of netting should be laid out. The main frames can be fixed before bed making. The frames can be made from 'L' angles or pipes. The netting can be done with GI wire of 16 gauge for length wise fixing and nylon threads for width. An increasing width of the meshes is used from bottom upwards. Generally, the bottom net is 7.5 x 7.5 cm, the subsequent nets are 10 x 10 cm, 12.5 x 12.5 cm and the upper net 15 x 15 cm. As the plants grow, the second, third and fourth layers can be put at 20 cm distance.

Nutrition

The application of nutrients in small doses but more frequently favours better growth and flower production. No nutrients are usually applied during the first three weeks after planting. Thereafter, nutrients are applied through fertigation. Carnation growers of the Nilgiris District of Tamil Nadu adopt the fertigation schedule detailed below, in general, with required modifications based on soil and water test results.

Irrigation

Rooted cuttings should be watered immediately after planting. Spraying of water through misting has to be done for at least two weeks. After 3 weeks of planting, drip irrigation has to be adopted. Drip irrigation system is being followed in carnation cultivation. Laterals are laid out every alternate row and emitters are spaced at 30 cm intervals. Fully grown carnation plants require 4-5 lit of water/m^2/day. It is advisable to check the irrigation water quality in terms of EC and pH. Stagnation of water should be avoided to minimize the incidence of diseases.

Weed Management

Weeding is a major problem during the early stages of crop growth. Weed growth will be suppressed when the plant covers the ground level. Only manual weeding is followed in carnation by using small hand hoes or forks. Soil loosening is combined with weeding for good aeration, water and nutrient percolation in to the root zone. It should be ensured that organic manures used should be free from weed seeds to avoid weeds.

Special Operations

Training

Training is a very important and continuous operation in carnation cultivation. This operation helps in keeping the plants within the specified area in the net to grow straight without bending at the bottom.

Pinching

Pinching is an important operation which helps to ensure maximum number as well as quality of flowers. Some cultivars do not require pinching but some cultivars require this operation to encourage branching. Pinching should be done approximately 3-4 weeks after planting. Three types of pinching are being adopted in carnation as detailed below.

Single Pinch

In this method, the top of the main or leader shoot is removed leaving 5 pairs of leaves from which 4-5 lateral shoots develop. These lateral shoots will produce flowers at the same time. This method is particularly applicable to cultivars which produce higher proportion of quality blooms like the hybrid standard carnations.

Pinch and-a-half

This method is followed to regulate the supply of flowers throughout the year. It involves the removal of main stem tip to induce 4-5 lateral shoots. When these lateral shoots develop 5-6 pairs of leaves, only half the number of the lateral shoots are pinched. This method provides steady supply of flowers but reduces the quantity of the first crop.

Double Pinch

In this method the main shoot is pinched once followed by pinching of all the lateral shoots arising from the first pinch when they are about 6-8 cm long or develop 5-6 pairs of leaves. This method produces larger number of flower bearing shoots but produce weak shoots and poor quality flowers. So this method is not commonly followed.

Deshooting

Unwanted secondary shoots on the flowering stems are removed when they are about 2-3 cm long.

Disbudding

Disbudding is practiced in standard and spray or miniature carnations. In standard carnations, disbudding is practiced for getting good quality flowers. The axillary/lateral buds are removed just after appearance without damaging the leaves and stems. Usually those axillary buds about six nodes below the terminal flower buds are removed to encourage the development of the main flower bud. In spray or miniature carnations, the main flower bud (terminal bud) is removed to encourage the lateral flower buds to develop.

Calyx Banding

Calyx splitting is a physiological disorder in carnation affecting the flower quality. This disorder can be minimized by calyx banding. Calyx banding is the practice of placing a rubber band or plastic tape around the calyx of the flower bud when it just begins to open.

Application of Growth Regulators

Growth regulators have been reported to significantly influence flower production and regulation in carnation. Two sprays of GA_3 @ 100 ppm each at first pinch and when axillary shoots are 8-10 cm long will produce early flowering and long stems. Growth retardants like chlormequat (CCC @ 0.25 per cent) and diaminozide (SADH @ 0.4 per cent) have been reported to promote flower initiation and increase flower yield. They also cause earlier flowering and improve flower quality by reducing calyx splitting. Application of Malic Hydrazide @ 500-1000 ppm also increases the number of flowers but delays flowering.

Harvesting

Most of the carnation cultivars will be ready for harvesting in about 105 to 120 days after planting. There are different stages of harvesting according the market demand. The harvesting stage is fixed depending on the type of carnation and market demand. Bud size and petal growth are used to judge the stage of harvesting.

Table 6.4: Harvesting Stages Based on Market Demand

Sl.No.	*Harvesting Stage*	*Target Market*
1.	Tight bud	This is for long distance markets stage. However, it is not practiced in commercial cultivation, since some of the flowers may not open at all after harvest.
2.	Paint brush	This stage is ideal for long distance market stage or for use after a couple of days.
3.	Semi-open	This is ideal for short distance market stage The flowers can be used in a day or two days after harvest.
4.	Open stage	This is ready to use harvest stage and not suitable for travel.

Table 6.5: Harvesting Stages Based on Type of Carnation

Type	*Harvesting Stage*
Standard	Carnations Paint brush stage, when petals have started to elongate outside the calyx.
Spray carnations	With two flowers open and rest of them showing colour.

Method of Harvesting

The best time of harvesting is morning. The best place of cutting stem is the area where leaves are well spaced and where at least two axillary shoots appear. The flowers are harvested by either snapping the stem off at a node or cutting off with a sharp knife or small shears. The cut given on the stem should be smooth to avoid injury to flower stem or to the mother plant. In the production period, flowers should be harvested at every two days interval. As soon as the flowers are harvested, the cut ends should be kept dipped in water or preservative solution (sodium hypochloride at 1 ml/10l of water).

Yield

Yield of carnations depends on type and cultivar, growing region and environment, planting time, plant density, pinching methods and also the period of harvest. In general, spray carnations produce more flowers than standard carnations. The number of flowers in standard carnations ranges between 8 and 12 flowers/ plant/year. In general 300-350 flowers/m^2 can be obtained from standard carnation, while 250 flowers/m^2 can be obtain from spray carnation.

Post-harvest Technology

Precooling

The harvested flowers are kept at a temperature of 4°C to 7°C soon after harvesting. Rapid precooling of flowers maintains quality as well as increases the longevity of flowers.

Grading

The flowers are graded to different classes according to their quality. Before grading the foliage at the bottom half of the stem should be removed (stripping). Standard carnations are graded based on length, strength of stem and flower size. The Society of American Florists suggests the following parameters to ensure quality. Flower and leaves should be bright, clean and firm. Flowers should have fairly tight petals at the centre. Flowers should be symmetrical in shape and size representative of the cultivar. Flowers should be free from calyx splitting. Flowers should be free of decay and mechanical damage. Flower stems should be free of lateral buds and suckers. Flower stems should be straight and have normal growth. Based on minimum flower diameter and stem length, the Society of American Florists has developed the following grades.

Table 6.6: Minimum Flower Diameter and Stem Length of Various Grades of Carnation

Grade	Parameters	
	Bud Dia (mm)	*Stem Length (cm)*
Fancy	60-75	55
Standard	45-60	45
Short	Blow 45	30

The European Economic Community has developed the following grades of carnation based on stem length.

Table 6.7: Minimum and Maximum Stem Length of Various Grades of Carnation

Description Code	*Minimum and Maximum Stem Length (cm)*
0	Less than 5 cm or flowers without stem
5	5-10
10	10-15
15	15-20
20	20-30
30	30-40
100	100-110
Till 120	>120

Bunching

After grading, the flowers are bunched in two layers with 20 flowers per bunch for standard carnation types and 10 stems per bunch (consisting of 35 opened or partially opened flowers) in each stem for spray carnation types.

Conditioning of Flowers

Carnation flowers are very sensitive to ethylene. Ethylene is a primary plant hormone involved in the senescence of cut carnation flowers. Conditioning the flowers after grading and bunching will increase the post harvest life of the flowers. After bunching the stem ends should be trimmed and placed in a preservative solution to extend the longevity of the flowers. The preservative solution is generally composed of a sugar, a germicide and growth hormones. Usually 8-hydroxyquinoline sulphate (8-HQS) or 8-hydroxyquinoline citrate (8-HQC) @ 200 ppm, silver thiosulphate @ 0.2 mM, cytokinin @ 50-100 ppm, sucrose @ 2.0 per cent and citric acid @ 50-100 ppm are used as flower preservatives. The pH of the preservative solution should be maintained at 4-5. The flowers should be placed in the solution at least for 4 hrs.

Packing

After conditioning the flowers are wrapped with polythene or polypropylene sleeves to protect them from mechanical damage and to improve the appearance. Normally perforated sleeves are used as wrapping material. The basal ends of the stems can be placed in absorbent cotton saturated with water and enclosed in waxed paper or aluminium foils. Different types of corrugated fibre board boxes are used for packaging of carnation flowers. The boxes must be strong enough to withstand the weight of at least eight full boxes placed on top of one another under conditions of high humidity. For long distance transport, telescopic style boxes made up of corrugated fibre board are recommended. Normally a box size of 122cm, 50cm and 30cm length, width and height, respectively is used. Standard carnations are packed with 24, 28 or 32 bunches per box according to the grade. Spray carnations are packed with 100 bunches per telescopic corrugated carton. All gaps inside the boxes should be filled with shredded paper. Sides of box should have vent holes with flap. Total vent size should be equal to 4-5 per cent of the area of the end wall of the box.

Storage and Precooling

After packing, the flowers should be cooled immediately to bring a relatively cool condition inside the packed boxes. Forced air cooling is normally followed. Cool air is forced into the boxes through the vents of the boxes. In this method, the forced air will bring down the flower temperature to the air temperature of the cold room within an hour. The optimum storage temperature may vary depending on stage of flower harvest and method of storage. Generally standard carnations harvested at bud stage are stored at 0°C, whereas, open flowers are stored at 3-4°C. The humidity must be maintained at a high level of 90-95 per cent. Refrigerated vans should be used for transporting the flowers.

Insect-Pests Management

Red Spider Mite

The red spider mite *(Tetranychus urticae)* is the most serious pest in carnation. Proper ventilation and watering are the most important factors to prevent the mite attack. Spraying Dicofol @ 2 ml/l or Wettable sulphur @ 5g/l or Propargite @ 2 ml/l during initial stages and Abamectin 1.9 EC @ 0.5ml/l of water with sufficient amount of surfactants during severe infestation stages will bring desired results.

Aphids

Aphids *(Myzus persicae)* suck the sap from the leaves of growing plants. Spraying the plants with Thiomethoxam 1 ml/l or acetamiprid 1 ml/l imidacloprid 17.8 SL 0.1 g/l will effectively control aphids.

Thrips

Thrips *(Thrips tabaci)* suck the sap from the leaves, causing them to turn yellow. Spraying Dimethoate 30 EC @ 1 ml/l or Fenitrothion 50 EC @ 3.5 ml/l or application of Aldicarb 10G @ 5 g/m^2 will control thrips.

Bud Borer

The caterpillars of the bud borer *(Helicoverpa armigera)* mostly affect the flower buds. Spraying of Nuvacron @ 3 ml/lit or Indocarp 14.5 SL @ 1 ml/l or Fenitrothion 50 EC at 1.5ml/l will control the bud borer.

Cut Worm

Young plants are often cut off at the surface of the ground. The worm is most prevalent in fields during June-July. Application of Diazinon @ 0.2 per cent to the soil prior to planting or spraying the plants with Sevin @ 0.15 per cent will control the cutworm.

Nematodes

Carnation is host to about 21 nematodes. The most important nematodes affecting carnation are *Cricomemoides curvatum*, *Cricomemoides xeroplex* and *Meloidogyne incognita*. The nematodes cause reduced root system, stunted shoot growth and reduction in the number of blooms. Nematodes can be eliminated by growing plants in fumigated soil. Application of Furadan, Aldicarp or Nemaphos @ 10g/m^2 controls nematodes in carnation.

Diseases

Carnations are attacked by a number of diseases which are caused by fungi, bacteria and viruses.

Fusarium Wilt

Fusarium wilt (*Fusarium oxysporum* f.sp. *dianthi*) is one of the most serious diseases in carnation. Poorly drained soil and abnormally high temperature are conducive for the development of this disease. The affected plants show foliage wilting, often only on a few branches followed by death. Rotting of the stem below ground level with internal brown streaking occurs. If the plants are pulled, they break off easily while the firm roots remain in the soil. The best control measures are soil sterilization or chemical fumigation of the soil, soil solarization using clear transparent polyethylene film (0.1 mm thick) for 30 days, use of pathogen free planting materials and general greenhouse sanitation. Drenching the soil with Benomyl (0.1 per cent) or Ridomil (0.2 per cent) at three month intervals starting from two weeks after planting and spraying with Bavistin @ 0.1 per cent will reduce the malady. Biological control with *Bacillus subtilis, Streptomyces sp, Trichoderma viride* and *Pseudomonas fluorescens* has also been reported to be effective against Fusarium wilt. These agents are generally applied @ 10g/m^2 at monthly intervals till the end of the first flush.

Alternaria Leaf Spot

Alternaria leaf spot *(Alternaria dianthi)* is a very common foliage disease in carnation. The pathogen causes spots on the leaves and stems and the affected leaves wither and die prematurely. The pathogen is present in the dead plant material and spreads by splashing water. Temperature above 23.8°C promotes growth of the pathogen. Foliar application of Dithane M-45 @ 0.2 per cent or Carbendazim @ 0.1 per cent controls the disease.

Bacterial Wilt

Bacterial wilt *(Pseudomonas caryophylli)* occurs in places where the night temperature is below 23.8°C. It is more common in older plants. The prominent symptom is wilting of one or more branches or the entire plant. Leaves of the affected plants look dull and greyish green and then turn yellow and finally die. Pseudomonas is a soil borne pathogen and hence soil sterilization and use of disease free planting material will minimize the disease incidence.

Bud Rot

This disease (*Rhizoctonia solani)* affects the plants at the soil line causing wilting and yellowing of foliage followed by death of plants. Butts sometimes show a brown discolouration and cracking just below soil level. The brown rot can extend up to the stem. Fluffy, light brown fungal hyphae can sometimes be observed on the surface of the rotting tissue. Early stage of the disease can be confused with Fusarium wilt but it differs in that no internal brown streaking is observed. The relative resistance of carnation plants to *Rhizoctonia solani* is increased by good air circulation and drainage and shallow planting of cuttings. The disease incidence can be reduced by drenching with Carbendazim @ 0.2 per cent before planting.

Rust

Early infections of rust *(Uromyces dianthii)* appear as pale green blister-like swellings which release reddish to dark brown powdery masses of spores. Pustules can be up to 10 cm in length and occur on stems, leaves and calyx. Severely infected leaves may turn yellow and die. The disease is common under warm humid conditions.

Keeping the foliage dry and regular preventive spraying of Carbendazim @ 0.1 per cent or Dithane M-45 @ 0.2 per cent can control the disease.

Grey Mould

Grey mould *(Botrytis cinerea)* is a storage disease that affects the petals. Initially a wet tan coloured blotch develops on petal tips which spread rapidly to produce a fluffy grey mould. This disease is favoured by high humidity. Reduced humidity and maintenance of good ventilation and hygiene practices and avoiding injuries to flower can minimize the disease. Benlate @ 0.1 per cent controls this disease.

Stem and Root Rot

The plants infected with stem and root rot *(Phytophthora* sp.) show withering and yellowing of foliage, leaf death, external browning of stems and internal browning at nodes. Wet conditions, over watering and badly drained soils favour development of this disease. The disease can be controlled by avoiding over watering and poorly drained soils. Drenching with Benomyl @ 2.5g/lit can be recommended for controlling the disease.

Fairy Ring Spot

The disease *(Heterosporium echinulatum)* causes fairy ring-like spots which coalesce, extend and merge, eventually destroying the leaves. Good aeration and spraying of Carbendazim @ 0.1 per cent or Dithane M-45 @ 0.2 per cent at 10 day intervals will reduce the disease infestation.

Viral Diseases

Carnation is subjected to many viral diseases. The most common ones are streak, mosaic, mottle, ring spot, etched ring and vein mottle. Use of virus free planting material raised through shoot tip culture can eliminate viruses. Vectors can be controlled by spraying of Thiomethoxan @ 0.1 per cent.

Physiological Disorders

Calyx Splitting

Calyx splitting is an important disorder in carnation which has been associated with many factors like genetic, environmental, nutritional and other cultural practices. The cultivars with short and broad calyx are more susceptible than the ones with long and narrow calyx. Irregular or fluctuating temperature during flowering also induces calyx splitting. Low temperature below 10°C leads to the development of an extra whorl of petals inside the calyx. The calyx unable to hold these extra growing petals splits up. Nutritional make up of plants also influence calyx splitting. Low nitrogen, high ammoniacal nitrogen or low boron levels enhance calyx splitting. Closer spacing has also been reported to encourage calyx splitting. Selection of cultivars that are less prone to calyx splitting, regulation of day (20-25°C) and night (12.5-15.5°C) temperatures and maintenance of optimal levels of nitrogen (25-40 ppm) and boron (20-25 ppm) in the growing medium can minimize this disorder. Spraying of borax @ 0.1 per cent at fortnightly intervals will reduce the disorder. Calyx splitting can be reduced by placing a rubber band around the calyx of the flower which has started opening.

Sleepiness

Sleepiness causes huge post harvest losses in cut carnation. It occurs due to exposure of flowers to ethylene or water stress. Also, the incidence of sleepiness has been found to be higher when the flowers are stored for a longer period or when they are exposed to high temperature. Spraying of STS 0.4 mM before harvesting the flowers will correct this disorder.

Grassiness

Grassiness refers to failure of plants to produce flowers. This is a genetic disorder which varies from variety to variety. Removal and destruction of affected plants is the only way of correcting this disorder.

Slabside

This disorder refers to uneven opening of flower buds resulting in the petals protruding on one side only, giving an asymmetrical and lopsided shape to the flower. It is common during cooler periods. This can be overcome by gradually increasing the temperature to optimum level.

Calyx Tip Die Back

Potassium deficiency and water stress cause tip dieback. The disorder commences with browning of the calyx tip and it progresses downwards damaging a major part of the calyx. This disorder is often followed by occurrence of secondary fungal infection which makes the flower unmarketable. Spraying of potassium chloride @ 5g/l two times at 10 days intervals and providing adequate water @ 4.5 l/m^2 can minimize this malady.

Internode Splitting

Splitting of internodes affect the quality of cut flowers. Splitting is due to boron deficiency. Application of borax @ 2g/m^2 will correct internode splitting.

7

Chrysanthemum

Chrysanthemum is one of the most widely cultivated garden flowers and ranks probably next to the rose in popularity. There is hardly any other garden flower which has such diverse and beautiful range of colour shades, widely different flower shapes and height range as chrysanthemum.

Classification

Chrysanthemum can be classified into several groups according to the size and shape of the flowers. The major groups are:

1. **Incurved:** The ray florets curve upwards and inwards towards the centre of the bloom to form a globular shape.
2. **Incurving:** Ray florets incurve loosely and irregularly and do not form a ball as above.
3. **Reflexed:** The florets are drooping in this type.
4. **Anemones:** The disc consists of disc florets and is hemispherical in form. The ray florets are either lingulate or quilled.

5. **Pompon:** Flowers are very small sized. Disc is concealed or absent.
6. **Singles:** Petals are arranged in one or not more than five rows with a prominent central disc.
7. **Korean:** Small, single or double flowers with a conspicuous central disc.
8. **Spoon:** Ray florets are tubular with a spatula or spoon like opening at the tips.
9. **Rayonnaetes:** The petals are quilled.
10. **Exhibition:** Outer florets are reflexed and inner incurved. Ray florets are generally twisted and irregularly overlap each other. A very attractive group.
11. **Spider:** Ray florets are elongated and tubular, curved and have hooks at the tip of the petals.
12. **Decorative:** Fully double flowers with flat petals and the central disc is generally absent.

Propagation

The chrysanthemum is propagated by two method:

1. Suckers
2. Cuttings

(1) Suckers

After flowering – Plants are beheaded to a height of about 15-20 cm from the ground level. New shoots start appearing in January-February round the base of the stool. These are separated and planted in 10 cm pots and kept in light shade for establishment.

(2) Cuttings

Plants are cut to ground level after flowering. New shoots arise from base as well as from axil of leaves. Young tip cuttings of 5-8 cm length are taken just above a node. The lower leaves are removed and basal portion of the cuttings are treated with some root promoting hormones like seradix before planting. The rooting media consists of 2 parts of loam, 1 part coarse sand and 1 part of leaf mould. The compost should be sterilized by stream or by 2 per cent solution of formalin. The time of raising of cuttings is June-July in north and east India. Cuttings are kept in partial shade for two weeks and irrigated 2-3 times in a day. The cuttings can be planted under mist for rapid rooting.

Potting

The suckers and cuttings are transferred to 10 cm pots immediately after rooting in a compost consisting of one part each of coarse sand, garden soil and leaf mould and traces of wood ash. The second potting is done two months after the first in the case of suckers raised during the February in a bigger pot of 15 cm. The potting mixture consists of a richer mixture containing one part sand, one part garden soil, two parts of leaf mould, a quarter part of wood ashes and two table spoonfuls of bonemeal. The final repotting is done in August in 25-30 cm pots. The potting mixture should be very porous and rich. The following mixture is recommended:

- ✰ Garden soil: 1 part
- ✰ Leaf mould: 2 parts
- ✰ F.Y.M.: 1 part
- ✰ Broken pieces of charcoal and bricks and wood ashes: ½ part
- ✰ Bonemeal: 2 table spoonfuls/pot.

Nutrition

The cuttings make vegetative growth till September in north and east India. Feeding of suckers with liquid manure should short from July and for cuttings from August. In other places, the feeding may be started 15 days after final repotting. The liquid manure is prepared by fermenting fresh cowdung and oil cake in a drum for 5-6 days (About 1-2 kg of each in 10 litres of water). It is diluted to tea colour and applied in the intervals of 7-10 days @ of 500 ml to 1 litre per pot. Feeding of weak fertilizer mixtures is also useful. It is suggested that a fertilizer solution containing 10 g of Potassium nitrate, 30 g of ammonium sulphate and 30 g of superphosphate dissolved in 10 litres of water should be applied twice during September at fortnightly intervals. About 500 ml of this solution is applied to each plant. In places other than north and east India, this solution can be applied one month after final potting. Over feeding should be avoided. Liquid manure should be continued till the buds are half open. If the pot is too vacant from the brim a top dressing covering 2.5 cm of the vacant space with the following compost is suggested during September.

1. Mustard/Neem cake ¼ part,
2. Cowdung/Horse dung ½ part,
3. Wood ash 1 part,
4. Garden loam 1 part.

Irrigation

Irrigation is very harmful. If the compost is porous enough, irrigation on alternate days during the hot summer of north India will not pose any problem. During rainy season, irrigation is regulated in such a way that the pots do not accumulate water. During the hot summer and after the rainy season, syringing to maintain turgidity and to clean the foliage dirt.

Staking, Stopping (Pinching) and Disbudding

Disbudding starts in October or as soon as flower buds appear. In chrysanthemum only one bud per stem is allowed to bloom and others are removed (disbudded). The ideal time for disbudding is when the buds surrounding the central one have developed 0.5 cm long pedicels. Too early disbudding may cause the injury to the central bud which is retained to bloom. Deshooting is also practiced from time to time by removing all side shoots before they attain the size of 2.5 cm. However, in singles' Koreans and sprays, no disbudding is done. For obtaining a single bloom per plant, no pinching is done but disbudding and dishooting are done. All suckers should also be removed from the base of the plant.

The methods of stopping (pinching) depends upon the nature of bloom one intends to obtain. If only one bloom per plant is retained, no stopping is needed. But if 3 or 6 stems are needed for the plant, stopping is resorted. The tip of the main stem measuring 3-5 cm is removed (Stopped or pinched) when the plants are 8-10 cm tall. This stopping will encourage the lateral shoots (or breaks) to develop from the leaf axils. Three strong laterals are retained and others removed. The laterals retained for flowering should preferably consist of one central stem and two on either side of it. The lateral are staked with strong split bamboo stakes inserted in the compost with a view to giving support and also to see that these are spread out from each other. The first crown bud develops at the end of each lateral which contains maximum number of ray florets and will give the largest bloom, though may not be the best bloom. This is retained and all other growth arising from the leaf axils is deshooted (removed). Sometimes the crown bud in the laterals is stopped to obtain the second crown bud which arises from the leaf axils in many cultivars. The second crown buds produce flowers of more intense colour, harder in texture and more symmetrical in growth. However, in most cultivars, the first crown bud produces the largest bloom.

Field Cultivation for Cutflowers

The small flowered double Korean type are also cultivated in the ground for making gardens and veni. Yellow and white coloured cultivars are preferred. The propagation is done from suckers. In south India, the flowering season is from July to January, whereas, in other parts, it is only during November to December or at the most upto the first week of January. The land is ploughed 2-3 times and 37-49 tons of F.Y.M. is added per hectare. Fertilizers at the rate of 61 kg N, 98 kg P_2O_5 and K_2O per ha may also be added. Some growers may also apply a top dressing of 61 kg of urea per ha at the time when flower buds appear. No pinching or disbudding is followed. Irrigation at the interval of 10-15 days depending on weather and weeding at regular interval are practiced. The yield of flowers is about 740-980 quintals. Some large flowering cultivars are also grown on a single stem or three stems to produce cut flowers for table decoration on a limited scale around cities.

Controlling Growth and Flowering

It is possible to regulate the growth and flowering by manipulating the day length and by chemicals. Both temperature and day length play an equal part in determining plant response of one over the other factor may be more. The rate of photosynthesis, respiration, absorption of water and minerals and transpiration are affected by temperature.

Controlling Growth by Day Length

If short day conditions are created by keeping the whole plant in a dark chamber or covering if effectively by black polythene sheets for 14 hours a day (from 5 PM to 7 AM), flowering time can be hastened by upto 30 days in many cultivars. Similarly blooming can be delayed till April in north India by treating the plants with long days. It is done by illuminating the plants with a combination of fluorescent tubes (40 w) and incandescent lamps (60 w) by hanging them alternatively 120 cm above the plants so that the plants do not get more than 9 h, of continuous dark period. Illumination during the middle of the night has been found to be more effective. Long day besides delaying blooming period initiate growth more vigorously resulting in increased size of plants and higher yield of flowers. Such L D treated plants need frequent pinching for branching. By giving treatment of L D and S D according to a schedule, it is possible to obtain flowering in every week starting from October and extending till April provided the temperature is optimum. Temperature between 10-30°C are tolerable. Temperature lower than 10°C inhibit flower bud initiation and above 30°C affect flower bud development.

Growth Regulation by Chemicals

B-Nine (0.25 per cent) helps to reduce the height of the plant if the spraying is done twice at 3 weeks interval just after pinching and 0.5 per cent if only one spraying is restored to B-nine imparts plant compactness, improves foliage colour and the proportion of different plant parts. GA_3 induces earlier flowering and higher yield in chrysanthemum.

Autumn Blooming Cultivars

At N. B. R. I., Lucknow some early flowering hybrids have been raised which openup the possibility of doubling the blooming time from 2-4 months.

Insect-Pests and Diseases

Insect-Pests

Aphid

Myzus persicae, Macrosiphoniella sanborni, Aphis gossypii and *Aphis fabae* are the aphids attacking chrysanthemum. Aphids are greenish to black dot-like insect which suck the sap from growing stem tips and under-surface of leaves, causing loss of vigour. They are also carriers of virus diseases and they concrete honey dew which forms a substraction for fungal growth. The control measures are spraying with tobacco soap decoction. Several insecticides, *viz.*, malathion and aldicarb have been used to control aphids.

Hairy Caterpillars (*Diaciisia obligera*)

It attacks the plants in rainy season and continuous till winter. The pest is easily recognized by the presence of hair on their body. They multiply first and have gregarious habit during early stage. As they ear up the leaves from surface, papery skeletons are left which dry up. Manual collection and destruction in early stages can cheek heavy infection. Spraying thiodon 35 EC or ecacelux 35 EC at 1.25 ml/l is recommended as a control measure.

Red Spider Mites

Mites (*Tetranychus urtecae*) look red dot-like bodies on undersurface of leaves causing white speeks in the early stages. They occur in hot season and damage leaves and buds which give a pale appearance. Orgadiazion compounds eg., Cyhexat in (25 g/100 l) are reported to give good control. Metasystox and kelthane are also effective.

Diseases

Fungal Diseases

Root Rot

Root rot in chrysanthemum is caused by *Pythium* and *Phytophthora spp.* Rooting of stem cuttings in rooting bed in warm moist conditions in most common but the fungus may also enter the estabilished plants through plant wounds. Sudden wilting of infected parts (roots, stem and leaves) in the diagnostic feature of Pythiun. Control measures are pre-planting sterliration of medium, good drainage and quick removal of infected cutting. Soil drench with thiram/captan mixture each at 2.5 g/m^2 prevents infection.

Phoma Root Rot

The disease is caused by *Phoma species*. The disease is more likely to occur in unsterilized soils. The symptoms are first noticed on lower leaves which become chlorotic and show signs of wilting. The disease progresses upwards causing necrotic spot and limping of leaves. Pre-planting soil dench with haban (140 g/100 l at 22 l/m^2) gives excellent control.

Foot Rot

The disease caused by *Rhizoctonia solani* is reported to be particularly severe in warm moist conditions. Infection of cuttings in rooting beds leads to mushy brown rot of stem and leaves. In contrast to *Phythium,* situated and conditions due to excessive water in soil reduces *Rhizoctonea* infection. Chemical control is possible by soil drench with thiram (150g/100 l).

Stem Rot and Wilt

The symptoms of this disease caused by *Fusarium oxysporum* occurs only after the appearance of flower buds or they show colour although infection might have taken place as early as in rooting bed. Stem near soil level becomes dark brown and dries. Lower leaves turn yellow and the plants wilt during the day and recover at night for some days to wilt permanently afterwards. Soil application of diathene M-45 effectively control this fungus in nursery stage.

Grey Mould

. The disease encited by *Botrytis cinerea* may attack. The plant at any stage from rooting bed to flowers in transit. Stem infection often starts from margin and proceeds towards centre and base showing occasionally semi-circular bands of infected tissue.

Flower infection starts with brown water soaked spots on leave spot. Spraying captan (100 g/100 l) dichlofluanid (50 g/100 l) have given good result.

Ascochyta Blight

It is caused by *ascochyta choysanthemi*. It is also called 'black rot' because of blackish lesions on stems and lower leaves which grow bigger to make irregular black blotches. On flower infection causes browning of petals on one side, spreading within the flower stalk which turns black and drops. Spraying with mancozeb (100 g/100 l) on plants gave good results.

Septoria Leaf Spot

In India, this is the most common and serious disease of chrysanthemum caused by *Septoria abesa*. Brown spots appear on leaf, grow in size and number in humid weather. Nudes turn yellow and eventually die welting of leaves facilities the infection. Foliar spray of Bavisten or benlate (0.1 per cent) effectively control this disease.

Powdery Mildew

A very common disease in several countries is caused by *Oidinum chrysanthemi*. It is characterized by white powdery patches of spores or upper surface of the leaves. Chemical treatment with iriforine (30 g/100 l) and carbendarin (25-30 g/100 l) is most effective.

White Rust

It is caused by *Puccinia horiana* and characterized by white pustules on underside of leaves. Application of mancozab has been recommended.

Verticillium Wilt

Verticillium wilt is also common malady in many countries including India. Symptoms occur only after flower buds appear when the plants wilt. Suddenly, although infection might have taken place much earlier. Branching with carbendazim (25-30 g/100 l) has been recommended for preventing its spreed.

Ray Speek

Dusky speeks on ray flowers are caused by either of the fungi, *Stemphylium floridamum* and *Alternaria chrysanthemi* or their combined attack. Application of zenchor maheb are used to control the disease.

Bacterial Disease

Bacterial Blight

Erwinia chryscenthemi causes blight disease in chrysanthemum and symptoms start with wilting of one or more branches during a sunny day with recovary at night. Stem tips turns brown, brittle and collapse. Stem become yellow with brownish streaks extending up to base. Control measures consist of soil sterilization, using disease free cutting and avoiding contamination. Use of streptomycin is advised only in propagation benches.

Viral Diseases

Some of viral diseases and control as given below:

Table 7.1: Symptoms and Control of Various Viral Diseases of Chrysanthemum

Viral	*Symptoms*	*Transmission Methods*	*Control*
Chrysanthemum stunt	Over all reduction in plant size, foliage pale in colour, flower may open prematurely, red and bronze, flowers after bleached.	During pinching	Use of cutting from virus free indexed stock
Tomato spalted wilt	Reduced growth, leaf may show ring and line pattern plants may show one sided development with necrotic streaks in stem/leaves.	Thrips, weeds and other plants	Control of thrips and weeds
Chrysanthemum mosaic	Plant sneak and small, flower may show in dark of florets	Aphids	Control of aphids
Chrysanthemum rosette	Yellow molting with mosaic pattern in sonae some cultivars enlarged veins wrinkled and rosette formation on terminal portion	Grafting	Use of cutting from virus-free indexed plants

8

Dahlia

Dahlia belongs to family compositae and having its origin in Mexico. It is a tuberous rooted half hardy herbaceous perennial with most gorgeously coloured flowers and are very popular in the Indian gardens. Dahlias are widely grown for garden display and indoor decoration. It has a wide range of flower colours and diversity in the form of the flowers and so can cater to the tastes of a large number of garden lovers. The height of dahlia plants varies from 30-180 cm depending upon the cultivar. The dahlia flower consists of a certain number of outer ray florets in which the male organs are modified into a strap-shaped petal, arranged round a central disc of bisexual florets. The ray florets in dahlia have all the flower colours whereas the disc florets are generally yellow. In the double flowered cultivars more of the male organs get converted into ray petals with proportionate reduction in the number of discs. The modern garden cultiars of dahlias have arisen as a result of crossing between various species and varieties mostly consisting of *Dahlia variabilis, Dahlia coccinea, Dahlia crocea, Dahlia imperialis, Dahlia jaurezii* and *Dahlia merckii*.

Classification

National Dahlia Society and Royal Horticultural society in England have classified the modern dahlias into 11 groups.

(1) Single flowered dahlias (2) Star dahlias (3) Anemone flowered dahlias (4) Collarette (5) Paeony-flowered (6) Decorative (7) Cactus (8) Double show and fancy dahlias (9) Pompon (10) Miscellaneous (11) Dwarf bedding.

Propagation

Dahlia is propagated by seeds and division of tubers. Seeds – sown in September-October in the plains and March-April in the hills. The seedlings are raised in the nursery and transplanted in beds after pricking only single types are propagated by seeds. Division of tuber is another method of propagation. Generally the old clumps will have 3 or more tubers attached to each stem and these should be separated with atleast one eye attached to it. The eye in the tuber is situated in the stem end. Tubers are planted during June in the plains while in hills, it is done during March-April. Tubers are planted 15 cm deep and 30-90 cm apart depending upon the class of dahlia. The large decorative types need more spacing than the dwarf bedding types. The cut end of the tuber should be treated with 0.2 per cent captan or 0.1 per cent benlate to prevent rotting. When only few plants are required this is the best method.

For raising large number of plants, cuttings are rooted. The cuttings are taken from the young green shoots, produced from the crown of the tubers when these are about 7.5-10 cm in length. The shoots should be cut as close to the crown as possible just below the first node above the tuber. This will encourage more growth from the secondary eyes which are as good as cuttings. The lower leaves in the cutting may be removed and should be treated with a root promoting hormones before planting. Cuttings are planted 5 cm apart in shallow boxes containing a porous mixture of coarse sand and leaf mould or peat or vermiculite alone. Under tropical conditions, cuttings planted in shade and sheltered from heavy rains may root easily. Rooted cuttings are pricked singally in 7.5 cm pots containing one part each of soil and leaf mould before final planting in pots or beds. In New Delhi and around, cuttings can be taken periodically from July to middle of September. Cuttings raised early may flower in December-January in New Delhi. Cuttings raised during December under protection flower at the end of February but the plants never grow more than 45 cm in height.

Culture

Soil

Dahlia can be grown in any type of soil but a medium textured soil of neutral pH or little acidic (pH 6.5) in reaction should be preferred.

Site

The site selected should be open with ample sunshine.

Manuring

About 4 kg of well rotten F.Y.M. or leaf mould per square metre of soil should be sufficient for most type of soils except for very poor or sandy soil where the quantity should be little more. Compost made from lawn mowings is also very good. Bonemeal at the rate of about 100 g per square metre area should also be added as basal dose. About a basketful of wood ash should be added per square metre at the time of bed preparation. One top dressing with cowdung or leaf mould at the time of the appearance of flower buds will be quite beneficial especially in poor soils. Vermicompost @ 2.5 ton, Azotobactor @ 2 kg and PSB @ 2 kg ha^{-1} gave significant highest plant height, plant spread, leave number, primary branches, first bud appearance, weight of flower, flower size, flower duration and weight of tuber.

Stopping and Thinning

When the plants have attained a height of about 25-30 cm and developed 4-6 pairs of healthy leaves, the plants are stopped. The giant and large decorative dahlias are allowed to have 3 side branches on light soil and 5 in rich soil. The alternate branches in successive internodes are removed. The process of deshooting is not followed for small decorative and cactus types unless these develop too many branches. In such events the weak branches should be removed. The side shoots should not be removed until these are about 10-15 cm in length.

Staking

Once the dahlias start growing vigorously, it will be necessary to provide strong stakes and to tie the plants to the stakes. It may be necessary even to stake the side branches. While staking and tying, no branch should be forced too far apart from its natural position. For one or two such odd branches one additional stake should be provided.

Disbudding

The central or crown bud is retained from blooming while other two are removed at pea stage. If the crown bud is damaged one side bud has to be retained in place of the central bud. Disbudding also includes deshooting *i.e.*, removing some of the side shoots arising from the leaf axils.

Irrigation

Dahlias are irrigated heavily at longer spells rather than frequent light irrigation. In cool and humid weather, it is irrigated once in 10 days. During hot spells the plants will need irrigation twice a week. In light soils frequency will be more than in light soils. Young plants require irrigation more frequently than established plants.

Pot Culture

The pot mixture consists of 1 part of garden soil, 1 part sand, 1 part leaf mould, 1 part manure and ¼ part of broken pieces of charcoal. About 2 table spoonful of bonemeal should be added to each pot. Pots of 25-30 cm dia are suitable for growing large dahlias. Liquid manure feeding is also helpful.

Frost

Dahlias are killed by frost. Therefore, they should be protected by thatching them or keeping the pots under a big tree.

Shading

Dahlia blooms raised for exhibition may be damaged by hot sun or dry winds and pouring rains. In dry and hot places the blooms may be shaded by cheese cloth, brown paper to prevent them from drying. The papers are fixed overframs supported by stakes.

Lifting and Storage of Tubers

After the flowering is over and plants turn yellow, the plants are headed back to 15-20 cm above ground level. Tubers are then lifted carefully with the help of a gardenfork taking care not to injure the crown. After lifting the tubers are cleaned by removing of soil and the tubers are then stored in cool dry rooms on shelves containing layers of sand. The tubers should be dusted with 0.2 per cent seven and sulphur to protect them from pests. Occasional spraying of water is necessary to prevent the tubers from shrinking.

Insect-Pests and Diseases

Insect Pests

Aphids

Aphids generally transmit viral diseases such as mosaic spotted wilt and affect the general growth of the plants. Common species of aphid observed on dahlia are *Myzus persicae*, *Braehycaudus helichoysi* and *Aphis fabae* and are controlled by spraying with 1.0 per cent metacid.

European Corn Borer (*Pyraussta nubilalis*)

The young larvae cause damage to shoot; flowers and leaves which get distorted and turn brown. The affected tips die completely after which the borers bore down through the stem causing wilting of all parts. The pest can be controlled by spraying with 1 per cent Dichlorovos.

Leaf Hoppers (*Empoasca fabae*)

Nymphs and aults suck the sap from the leaves. The infected portion turns pale yellow, later it becomes brown and brittle. Spraying with 1 per cent Malathion is effective to control leaf hopper.

Beetes (*Oxycetonia albopunctata*)

The beetes faced on flowers. Spray with carbaryl (0.1 per cent) controls this pest.

Thrips

Several species of thrips (*Megaleurothrips distalis*, *Mocrocephalothrops abdominalis* and *Thoips nawaiiensis* enfest flowers. They feed on the juicy portion of leaves, causing a white speeking. Spray with 1 per cent Endosulphan is effective to control the pest.

Red Spider Mite (*Titranychees urticae*)

Red spider mites are serious pests of dahlia. They cause damage to the foliage by feeding on the undersides of leaves, causing a while sparkling. Spray with 1 per cent Endosulphan is effective to control the pest.

Diseases

Fungal Diseases

Powdery Mildew

Powdery mildew (*Erysiphe chicoracearum*) is one of the most important fungal diseases of dahlia. It is characterized by the development of white or gray powdery mass on the upper surface of the leaves. The disease is most prevalent in the early stages of plant growth and reappears at the end of the flowering season. The disease is mainly responsible for the rotting of seed heads sprays with Morestan (0.1 per cent) Bavisten, Benlate (0.1 per cent) or wettable sulphar (0.2 per cent) can effectively control this disease.

Bud and Flower Blight

This disease is caused by the fungus *Botoytis cinerea*. It occurs in cool and moist climates and produces water-soaked spots on buds and flowers. The infected areas get covered with gray-coloured mycelium and spore growth, which ultimately become feeded and with brown in colour. The disease can be controlled by spraying with Zinebs or Captain (0.2 per cent).

Stem Rot

The fungii (*Sclerotinia sclerotiorum*) attacts the main stem and branches near the base of the plant in nursery as well as after transplanting. The effectd parts firstly develop light brown water soaked patches that rot in cloudy weather and are covered by white mycelia growth. So the plants ultimately weather and die. Overcrowding of plants should be avoided. Soil may be sterlised with steam or chemicals like formaldehyde.

Wilt

Fusarium oxysporum, Fusarium dahlia and *Verticillum dahlia* are responsible for wilt. The plants wilt and die because of the blockage or destruction of the water conductor tissueses due to the toxins produced by the fungus. Wilted plants should be destroyed and only disease-free planting materials should be used in the propagation.

Smut

The smut caused by *Entylome dahle* is a destructive disease occurring in humid areas. The disease is characterized by the development of circular, light green spots which appear on both sides of leaves. Later, the centre of the spots turns brown and is sorounded by a yellow margin. In severe cases, the leaves show shot hole symptems and dry up. Diahane Z-78 (0.2 per cent) or Dithane M-45 (0.2 per cent) sprayed at forthnightly intervals are effective for controlling the disease. The use of Blitox-50 (0.2 per cent) and Benlate (0.1 per cent) was also suggested.

Leaf Spot

Leaf spot caused by *Cercospora grandissima* and *Cercospora dahlicola*. Infection is distinct in the form of angular to irregular dark brown spots which coalesce, forming large patches. The disease is serve in humid climate. Spraying with Bavistin (0.1 per cent) or Zineb (0.2 per cent) is effective. Diseased leaves and plants should be removed and burnt.

Bacterial Diseases

Crown Gall

Crown gall caused by *Agrobacterium tumefaciens* is characterized by the development of abnormal tissues at the base of the plant and roots. Infected plants become stunted and the shoots spindly. The disease can be prevented by dipping the roots in streptomycin solution during planting. Roots and crowns of plants with tumors are destroyed and crop rotation should be followed.

Bacterial Wilt

The disease is caused by the bacterium *Pseudomonas solanacearum* and the infected plants droop and wilt suddenly. Stems at the collar region rot and ooze yellow slimy mass of bacteria after cross section. Dipping of the roots in antibiotic solution and crop rotation are the only remediae measures of this disease.

Bacteriosis

The disease caused by *Erwinia cytolice*, was first recorded on Dahleas grown at the New York Botanical Garden. Stem become brown and rot. Pre-sowing dipping of roots in antibiotic solutions and crop rotation are the only control measures for this disease.

Viral Diseases

Mosaic

This is the most serious virus diseases of dahlia. The leaves show moltling and vein-banding symptoms. The virus is sap transmissible and creates general mosaic or pale yellow spliting. It is also transmitted by *Aphis gossypii*. ELISA techniques have been used to detect the virus. Control of vectors by spraying with insecticides is one of the preventive measures for controling the disease. Meristem tip culture was found to be a very effective method to eliminate the disease from the enfected plants.

Spolted Wilt

The disease, caused by *Lethum australiense* is known as ring spot and appears as conspicuous rings of light green or yellow tissues. Ring patterns appears later. Control of vectors by spraying with insecticides and meristem tip culture are the only remedial measures for the disease.

Cucumber Mosaic Virus (CMV)

It produces mosaic symptoms on the foliage. The virus has been deteched through dot-blot hybridization in dahlia. Control of vactors by spraying with insecticide at 15-day intervals and follow up of meristem tip culture are effective for this disease.

9

Gerbera

Gerbera is native of tropical regions of South America, Africa and Asia. The first scientific description of a *Gerbera* was made by J.D. Hooker in Curtis's Botanical Magazine in 1889 when he described *Gerbera jamesonii,* a South African species also known as Transvaal daisy or Barberton Daisy. *Gerbera* is also commonly known as the African Daisy. *Gerbera jamesonii,* native to Natal and Transvaal and *Gerbera viridifolia* from Cape, were crossed by Lynch in 1887 after both these species were introduced to Cambridge Botanic Garden and the hybrid was named as *Gerbera cantabrigensis. Gerbera* species bear a large capitulum with striking, two-lipped ray florets in yellow, orange, white, pink or red colours. The capitulum, which has the appearance of a single flower, is actually composed of hundreds of individual flowers. The morphology of the flowers varies depending on their position in the capitulum. The flower heads can be as small as 7 cm in diameter or up to 12 cm. *Gerbera* is very popular and widely used as a decorative garden plant or as cut flowers. The domesticated cultivars are mostly a result of a cross between *Gerbera jamesonii* and another South African species, *Gerbera*

*viridifolia.*The cross is known as *Gerbera hybrida.* Thousands of cultivars exist. They vary greatly in shape and size. Colours include white, yellow, orange, red and pink. The centre of the flower is sometimes black. Often the same flower can have petals of several different colours. It is an important commercial flower grown throughout the world in a wide range of climatic conditions. This is ideal for beds, borders, pots and rock gardens. The flowers are of various colours and suit very well in different floral arrangements. The cut blooms when placed in water last for a long time.

Morphology

Native to South African and Asiatic regions. They mostly inhabit temperate and mountainous regions. In India, they are distributed in the temperate Himalayas from Kashmir to Nepal at altitudes of 1300 to 3200 metres. Gerbera belongs to the family Compositae or Asteraceae. Plants are stemless and tender perennial herbs. Leaves radical, petioled, lanceolate, deeply-lobed, sometimes leathery, narrower at the base and wider at top and are arranged in a rosette at the base. The foliage in some species are entire and have lighter under-surface. Flower heads solitary, many flowered, the conspicuous rays in 1 or 2 rows, those of the inner row, when present, very short and sub-tubular and 2-lipped. The daisy-like flowers are in wide range of colours including yellow, orange, cream-white, pink, brick-red, scarlet, salmon, maroon, terracotta and various other intermediate shades. The double cultivars sometimes have bicoloured flowers which are very attractive. The flower stalks are long, thin and leafless. Achenes beaked and with pappus.

Based on flower heads, they may be grouped into single, semidouble and double cultivars.

Species

Gerbera ambigua, Gerbera aurantiaca, Gerbera bojeri, Gerbera bonatiana, Gerbera connata, Gerbera cordata, Gerbera crocea, Gerbera curvisquama, Gerbera delavayi, Gerbera diversifolia, Gerbera elliptica, Gerbera emirnensis, Gerbera galpinii, Gerbera gossypina, Gerbera hypochaeridoides, Gerbera jamesonii, Gerbera kunzeana, Gerbera latiligulata, Gerbera leandrii, Gerbera leiocarpa, Gerbera leucothrix, Gerbera lijiangensis, Gerbera linnaei, Gerbera macrocephala.

Cultivars

- ☆ TNAU Varieties: YCD-1, YCD-2
- ☆ Red: Ruby Red, Sangria, Dusty, Fredorella, Vesta.
- ☆ Yellow: Doni, Supernova, Mammut, Talasa, Uranus, Fredeking, Nadja.
- ☆ Rose: Rosalin, Salvadore.
- ☆ Pink: Pink Elegance, Marmara, Esmara, Terra-queen, Valentine, Fredaisy.
- ☆ Orange: Carrera, Goliath, Marasol.
- ☆ Cream: Farida, Dalma, Snow Flake, Winter Queen.

Cream Clemetine (Cream), Maron and Clementine (Orange) are reported to be a very high-yielding cultivar producing 434 flowers/m^2, Flamingo (Pale Rose), Delphi

(White), Labalga (lilac) are of commercial importance throughout the world. Amber, Romilda, Rozamunde, Hildegard, Alexis, Bilitis, Anke, Appelbloesem, Marleen, Sympathie, Salmrosa and Pascal are some other promising cutivars.

Propagation

Both sexual and asexual methods are common.

Seed Propagation

The advantages of seed over vegetative propagation include economy, simplicity and absence of pest and disease transmission. Seeds can be produced by both indoor and outdoor plants. Seeds are sown immediately after harvesting as they remain viable for 15 days. After germination, seedlings are picked at 2-4 leaf stage and are ready for transplanting in about 5-6 weeks. Seed propagation, however, is not always satisfactory.

Vegetative Propagation

Division

This method is the most common method involving dividing large clumps into smaller units, practised in June, when the plants may be set out in the field. In greenhouses they should be benched in the fall. Before transplanting, the roots and leaves of the suckers are trimmed, keeping the central shoot intact. Care should be taken that soil does not cover up the central growing point while setting the suckers in a new bed.

Cutting

Suitable plants are kept without water for 3 weeks, roots pruned, planted in peat and held at 25-30°C with 80 per cent relative humidity. The buds in the axils of the leaves are detached and rooted in rooting medium. They are ready for transplanting in 2 or 3 months. Approximately 40-50 plants can be produced in 2-3 months from a single mother plant. Young stem cuttings produce roots and shoots much easily and quickly under intermittent mist.

Micro-propagation

This is suitable for commercial propagation of large number of plants. It involves rapid multiplication of explants by repeated sub-culturing and preparation of divisions for transfer to soil. Shoot tips, inflorescence buds, flower heads, capitulum and mid-ribs have been employed as explants for micro-propagation.

Soil and Climate

Soil

A well-drained, rich, light, neutral or slightly alkaline soil is most suitable for gerbera production. In soils with good structure and drainage the root system reaches very deep and drainage appeared to be the most important factor. In soils with a pH range of 5 to 7.2, gerbera produced significantly more flowers and slightly longer flower stems than when grown at a soil reaction below pH 5.

Climate

In tropical and sub-tropical climate, gerberas are grown in the open but in temperate climate they are protected from frost and cultivated in greenhouse. They like sunny situation in cool weather but during summer months they should be lightly shaded if left out in beds. Poor light during winter adversely affects the flower production. A night temperature of 12°C was found to be optimum.

Cultivation

Field Preparation

Soil fumigation with Formaldehyde (100ml in 5l/m^2) or Dazomet (30g/m^2) is recommended to control soil borne pathogens (*Phytophthora, Fusarium* and *Pythium*). Raised beds of 1-2m width and 30cm height are prepared. Growing media consisting of FYM: sand: cocopeat/paddy husk (2:1:1) is ideal. Before starting gerbera cultivation, disinfection of the soil is absolutely necessary to minimize the infestation of soil borne pathogens like *Phytophthora, Fusarium* and *Pythium* which could otherwise destroy the crop completely. The beds should be drenched/fumigated with 2 per cent formaldehyde (100 ml formalin in 5 litres of water/m^2 area) or methyl bromide (70 g/m^2) and then covered with a plastic sheet for a minimum period of 2 to 3 days. The beds should be subsequently watered thoroughly to drain the chemicals before planting. Well developed tissue culture plants having 4-6 leaves can be planted firmly without burying the crown.

Planting

The plants should be set with the crowns well above the soil surface. In June, the clumps should be dried off gradually in the beds or removed to cold frames or field and replanted in the fall. The highest total cut-flower yield can be produced by planting in mid May to mid July. A good winter crop can also be obtained in protected cultivation in October to March. Spacing of about 30 x 20-30 cm (large number of flowers/m^2) for annual crop and 30 x 40 cm (less yield) for perennial crop, is suitable. Growing of gerbera in raised beds improves drainage and aeration.

Manures and Fertilizers

Gerbera requires plenty of organic matter in the soil for proper growth and flowering. An application of 7.5 kg FYM/m^2 and 12:8:25 g NPK/m^2/week is recommended. For greenhouse cultivation, 80 per cent of the recommended dose (12:8:25 g NPK/m^2/week) of fertilizers through fertigation along with foliar application of micro nutrients (multiplex 2 ml/l of water) at weekly interval is recommended.Fertigation is adopted from 3rd week after planting as per the following schedule.

Fertilizer	*Quantity (g/500m²)*
A tank (Monday, Wednesday, Friday)	
Calcium Nitrate	700
Pottasium Nitrate (13:0:46)	400
Fe EDTA/sulphate	20
B tank (Tuesday, Thursday, Saturday)	
Mono Ammonium Phosphate (12:61:0)	300
Sulphate of Potash (0:0:50)	700
Magnesium Sulphate	700
Manganese Sulphate	5
Zinc Sulphate	3
Copper Sulphate	3
Molybdenum (Sodium Molybdate)	1
Boron (Borax)	3

Irrigation

Gerbera needs thorough irrigation instead of light sprinkling at frequent intervals. Drip irrigation is done once in 2 – 3 days @ 3.75 litre/drip/plant for 15 – 20 minutes. Average water requirement is about 500 – 700 ml/day/plant. However, water-logging should be avoided as it is harmful to plants. Irrigation is provided in monthly intervals, twice in summer and once in winter.

Growing in Pots

Gerberas are also suitable for pot culture. A pot compost which is porous, rich in organic matter and with high moisture retentive capacity suits the requirement of gerbera. Gerberas can be grown better in peat or a compost mixture of 90 per cent pine bark and 10 per cent birch bark or pine bark and sphagnum peat moss. It is important to point out that the quality of flowers grown in pots declines considerably after the first year. So it would be beneficial to repot the plants dividing clumps immediately after flowering is over.

Harvesting

Flowers are generally cut when the outer two rows of disc florets are perpendicular to the stalk. In some cultivars, especially the ones that close at night can be harvested bit later. The stage of harvest is critical as the flowers should not be cut before the outer row of flowers show pollen. They respond very well to re-cutting of stem before placing in water or preservative solution or at all stages of marketing. Shortage of water as a result of vessel blocking or microbial growth causes drooping or wilting of flowers. Water is absorbed through the medullary cavities that are visible at the cut-surface. Therefore, it is beneficial to prick the stalk 10 cm below the flower head to allow the air to escape from the cavity. Gerberas are not suitable for long-term storage as the flowers lose 40 per cent of their vase-life even after 7-day storage. For short-term storage, the optimum temperature is 1.7°C, however, can be stored at 4°C for up to 8 days. The flower heads are packed in insulated boxes to protect them from cold or freezing temperature. Plastic-coated metal grids measuring 50 x 70 cm with a mesh size of 2 x2 cm can be used to pack gerbera flower heads. Flower heads supported by the grid which is suspended above a plastic tray measuring 48 x 70 x 30 cm, at a height which can be adjusted according to the length of flower stalk so that the flower ends of the stalk are immersed in water to a depth of 15 cm in the tray.

Yield

The crop yields 2 stems/plant/month. Harvest starts from 3rd month of planting and continued up to two years. Under open condition, 130 -160 flowers/m^2/year and under greenhouse condition, 175 - 200 flowers/m^2/year can be obtained.

Post-harvest Handling

Harvesting is done when outer 2-3 rows of disc florets are perpendicular to the stalk. The heel for the stalk should be cut about 2-3 cm above the base and kept in fresh chlorinated water. Flowers should be graded and sorted out in uniform batches. Flowers packed individually in poly puches and then put in to carton boxes in two layers.

Disorders

An abnormality characterized by numerous leaves, short petioles and small laminae which gives some cultivars of gerbera a bushy appearance known as bushiness. Nodes are not clearly distinguished and no internode elongation is seen.

Stem Break

It is a common post harvest disorder in cut gerberas. This is mainly caused by water imbalances. It could be ethylene controlled and associated with early senescence caused by water stress.

Yellowing and Purple Margin

Nitrogen deficiency causes yellowing and early senescence of leaves. Phosphorus deficiency causes pale yellow colour with purple margin. Increase in levels of nitrogen and phosphorus were found to promote development of suckers and improve flowering in gerbera.

Grading

Based on stem length and diameter, flowers are graded in A, B, C and D.

Vase-life

Proper harvesting, post-harvest handling and use of floral preservatives improve the keeping quality of flowers. For good vase-life, cut-flowers should be placed immediately after harvest in fresh water. The addition of chrysal-VB solution to the water increases vase-life. Gerberas are the only cut-flowers that are not damaged by adding chlorine to water. At the correct dosage, it will kill bacteria and prolong vase-life. Cutting the base of stem of the cut-flowers just before placing in water and giving a 24 hours dry storage to stimulate transport conditions improves the keeping quality. Several preservatives have also been found very effective in prolonging the vase-life of cut-flowers. Immersing cut-flowers for 24 hours in a solution containing 200 mg/l hydroxyl quinoline sulphate (HQS), 50 g/l silver nitrate and 5 per cent sucrose improves their vase-life by 4-5 days.

After Cultivation

1. Hand weeding is done whenever necessary.
2. Remove the flower buds up to 2 months and then allow for flowering.
3. Rake the soil once in 15 days to facilitate easy absorption of water, fertilizer and to provide air to the roots.
4. Remove older leaves to facilitate new leaf growth and good sanitation.

Pests and Diseases

Pests

Among different insect-pests, whiteflies, aphids, leaf miner, spider mites, caterpillars and root-knot nematodes are common attacking gerbera plants.

Whitefly

In greenhouse, gerbera is a subject to heavy infestation of whitefly (*Trialeurodes vaporariorum*), a sucking insect. Control measures include biological control with *Encarsia formosa* and the use of Aerosols, smokes and Parathion and Malathion sprays. An integrated control with the use of yellow-orange plexiglass sticky traps was also suggested.

Leaf Miner

The leaf mining fly, *Lirioyza trifolii,* is a serious pest of gerbera. Another species, *Lirioyza soncho,* was also noticed. Adults of these small winged insects lay eggs on the leaf. The larvae bore into the leaf and make irregularly shaped tunnels or blotches which are generally light yellowish tan to brown in colour. As these larvae mature they fold the leaf together with threads and feed on the inner surface. Application of insecticide, Ambush (Permethrin) @ 50 ml per 1000 m^2 by 'swingfog' or 'pulsfog' atomizers repeatedly if needed, is recommended. Dimethoate @ 0.1 per cent also gives good control.

Mites

Greenhouse-grown gerberas are often attacked by *Hemitarsonemus latus* and *Steneotarsonemus pallidus*. The development of leaves and flower buds was adversely affected and the flowers were malformed and unsaleable. Good control can be obtained by spraying the plants with 20 per cent Endrin @ 200 ml/100 litre of water.

Aphids

This insect infests young leaves and buds and causes injury by sucking the sap which results in distortion of tissue. *Myzus persicae* and *Aphis fabae* are the main insects that cause damage. Spraying with 0.1 per cent Dichlorvos or fumigation with Sulfotep smoke cartridges at 200 m^3 are recommended for their control. Among other insect-pests that occasionally infest gerbera are *Heliothis armigera, Mamestra brassicae* and *Spodoptera liltoralis.*

Nematodes

Root-knot nematode is occasionally serious in gerbera. Temik (Aldicarb) granules can be used for controlling soil infestations of *Meloidogyne spp.* under glass. This nematicide considerably reduces the nematode population.

Diseases

Diseases can be a major factor limiting gerbera production. Important diseases and their control measures are given below:

Fungal Diseases

Root Rot

It is a soil-borne disease caused by *Pythium irregularae* and *Rhizoctonia solani.* The infection results in stunted growth and ultimately drying of entire plants. *Rhizoctonia solani* causes more losses than *Pythium irregulareae* and can attack older plants. Soil sterilization gives good control of the disease.

Foot Rot

This disease is also caused by soil borne pathogens and *Phytophthora cryptogea* is very common attacking gerbera. Foot rot of *Gerbera jamesonii* is caused by *Fusarium oxysporum.* The short stems blacken and rot and the leaves and flowers die. Soil sterilization with vapam at 100 ml/m^2 was found very effective to control this disease. Fungicide like Prothiocarb @ 0.15 per cent or Topsin @0.05 per cent gave better control of *Pythium cryptogea.* Soil drenching with 0.4-1.6 g Metalaxyl + 0.8-3.2 g Folpet/m^2 also gave better results.

Sclerotium Rot

Sclerotiium rolfsii was identified as the cause of this serious fungal disease affecting the entire above ground parts of the plants. The incidence may be as high as 90 per cent during August and September when the atmospheric relative humidity was high and the temperature ranged between 30° and 34°C.

Blight

This disease, also known as gray mould, is caused by *Botrytis cinerea* which kills young growing tissues. Flower heads of gerbera growing in humid conditions under glass develop very small black spots on the ray florets. An evidence of a distinct blocking reaction of the host cells, restricting the fungal mycelium to very small areas of the petals is also noted sometimes. Deep planting, bad drainage and poor ventilation pre-dispose plants to infection and should be avoided. Watering should be discontinued from mid-October until the end of March. Infected parts should be removed and destroyed. Spraying of 0.7 per cent Captan, 1 per cent TMTD or 0.7 per cent Euparen at 8 to 10 days interval is suggested to control the disease. Benlate 0.1 per cent or Thiram 0.1 per cent treatment also controls the infection.

Powdery Mildew

Two fungi, *Erysiphe cichoracearum* and *Oidium erysiphoides* f. sp. *gerbera*, producing a white powdery coating on the foliage, have been reported on gerbera. Mildex (Dinitro capryl phenyl crotonate) proved to be the most effective fungicide to control the disease without plant injury. Spray with wettable sulphur or Karathane controls *Erysiphe cichoracearum* well and to prevent *Oidium erysiphoides* f. sp. *gerbera*, the relative humidity should be reduced, especially in spring and autumn, old leaves should be removed and the house ventilation is provided. Good control is obtained by applying contact fungicides, elemental sulphur or Dinocap or the systemic fungicides like Benomyl 0.1 per cent or Topsin M 0.08 per cent, before flowering. On flowering plants, only Benomyl or Pyrazophos should be used.

Leaf Spot

It is also a common disease caused by *Phyllosticta gerberae, Alternaria* spp. and others. In greenhouse and field tests, Bordeaux mixture gave little control. *Alternaria* leaf spot, but Captan was found promising. Spraying of 1 per cent Bordeaux mixture, Maneb 0.5-0.7 per cent, Zineb 0.5 per cent, Polyram 0.4 per cent or Ziram 0.4 per cent is also recommended to control the disease.

Downy Mildew

This disease for the first time had been reported from Poland and the causal organism was identified to be *Bremia lactucae.* The disease can be effectively controlled by Ridomil 25 WP (Metalaxyl) @ 0.1-0.15 per cent by spraying 3-6 times at fortnightly interval or by soil drenching @ 0.7-1.0 g/plant twice at 3-4 weeks interval. Dithane M-45 (Mancozeb) or Captafol or Captan @ 0.3 per cent used 8 times at 9-10 days interval also proves useful in controlling the disease.

Some other fungi have also been reported to attack gerbera plants like *Scleotinia sclerotiorum and Verticillium dahliae* causing severe stunting and wilting of plants. Strict hygiene and careful choice of preceeding crops and fumigation with Methyl bromide, chloropicrin or mixtures (70:30) of the two is recommended for control of the diseases.

Bacterial Disease

Bacterial Blight

A several leaf blight disease which limited the gerbera production was characterized by small to large, circular to irregular, brownish black leaf spots with or without concentric rings. Large brown areas, extending from the margins and narrowing as they reached the mid-veins of the leaf, often occurs.

Viral Diseases

Tobacco Rattle Virus

This disease was first reported from Florida. Yellow or black annulated ring spots were observed on the foliage of field grown gerbera. Soil steaming before each crop is recommended to destroy the nematode vectors such as *Trichodorus* spp. and prevent the disease.

Mosaic

The mottling caused by gerbera mosaic virus is reported to be transmitted by grafting.

10

Gladiolus

Gladiolus is a genus of perennial cormous flowering plants in the iris family Iridaceae. It is sometimes called the 'Sword lily' but usually by its generic name. The genus occurs in Asia, Mediterranean Europe, South Africa and tropical Africa. The center of diversity is in the Cape Floristic Region. The genera *Acidanthera*, *Anomalesia*, *Homoglossum* and *Oenostachys*, formerly considered distinct, are now included in *Gladiolus*. The genus *Gladiolus* contains about 260 species, of which 250 are native to sub-Saharan Africa mostly South Africa. About 10 species are native to Eurasia. There are 160 species of *Gladiolus* endemic in southern Africa and 76 in tropical Africa. The flowers of unmodified wild species vary from very small to perhaps 40 mm across and inflorescences bearing anything from one to several flowers. The spectacular giant flower spikes in commerce are the products of centuries of hybridisation, selection and perhaps more drastic manipulation. Gladioli are half-hardy in temperate climates. They grow from rounded, symmetrical corms that are enveloped in several layers of brownish, fibrous tunics.Their stems are generally unbranched, producing 1 to 9 narrow, sword-shaped, longitudinal grooved leaves, enclosed in a sheath. The lowest leaf is shortened to a cataphyll. The leaf blades can be plane or cruciform in cross section. The flower spikes are large and one-sided, with secund, bisexual flowers, each subtended by 2 leathery, green bracts. The sepals and the petals are almost identical in appearance and are termed tepals. They are united at their base into a tube-shaped structure. The dorsal tepal is the largest, arching over the three stamens. The outer three tepals are narrower. The perianth is funnel-shaped, with the stamens attached to its base. The style has three filiform, spoon-shaped branches each expanding towards the apex.The ovary is 3-locular with oblong or globose capsules containing many, winged brown, longitudinally dehiscent seeds. In their center must be noticeable the specific pellet-like structure which is the real seed without the fine coat. In some seeds this feature is wrinkled with black colour. These seeds are unable to germinate.These flowers are variously colored, pink to reddish or light purple with white, contrasting markings, or white to cream or orange to red.The South African species were originally pollinated by long-tongued anthrophorine bees but some changes in the pollination system have occurred allowing pollination by sunbirds noctuid and Hawk-moths, long-tongued flies and several others. In the temperate zones of Europe many of the hybrid large flowering sorts of gladiolus can be pollinated by small well known wasps. Actually, they are not very good pollinators because of the large flowers of the plants and the small size of the wasps. Another insect in this zone which can try some of the nectar of the gladioli is the best-known European Hawk-moth *Macroglossum stellatarum* which usually pollinates many popular garden flowers like *Petunia, Zinnia, Dianthus* and others.Gladioli are used as food plants by the larvae of some Lepidoptera species including the Large Yellow Underwing.Gladioli have been extensively hybridized and a wide range of ornamental flower colours are available from the many varieties. The main hybrid groups have been obtained by crossing between four or five species followed by selection: Grandiflorus, Primulines and Nanus. They make very good cut flowers. The majority of the species in this genus are diploid with 30 chromosomes but the Grandiflora hybrids are tetraploid and possess 60 chromosomes. This is because the main parental species of these hybrids is *Gladiolus dalenii* which is also tetraploid and includes a wide range of varieties.

Gladiolus is an important cut flower crop, grown commercially in many parts of the world. It has gained popularity owing to its beauty, attractive colours, various sizes and shapes of flowers, variable spike length and long vase life. Gladiolus, with their beautiful spikes produce flowers from October to March in the plains and from June to September in hills of India. The most important gladiolus growing countries are Holland, Germany and the United States of America. This plant is also grown commercially in India for production of cut flowers and corms. Gladiolus is commercially cultivated in West Bengal, Himachal Pradesh, Sikkim, Karnataka, U.P., Tamil Nadu, Punjab and Delhi. There are over 30,000 varities of Gladiolus.

Classification

A. Horticultural Classification

Large Flowered Hybrids

The large flowered gladiolus is more suited for garden display than floral decorations with triangular overlapping florets. The flower spikes grow to a height of 120 to 150 cm, are strong and erect, with florets of 10 to 15 cm across closely arranged triangular and symmetrical flowers and flowers late in the season.

Primulinus

Primulinus gladiolus has slender spikes with separate wide florets arranged in a zig-zag pattern. They bear hooded flowers which are smaller in size. The spike grows to a height of 100 cm and bears florets of 5 to 10 cm across and a mid- season flowering type.

Miniatures

Miniature gladiolus spike grows to a height of 60 to 100 cm, with florets of 5 to 7.5 cm across, good for cutting and early flowering type. They include butterfly hybrids which possess spikes up to 36 cm in length. Edges of the petals are often frilled and ruffled with individual florets about 6 cm in diameter, having most distinctive throat markings.

Peacock Hybrids

These are good for cutting, dwarf in height, multi coloured sorts with reflaxed petals.

Star Flowered

These types bear flat star-like flowers.

B. On the Basis of Spike Length and Floret Number

Grade	*Spike Length (cm)*	*Minimum Floret Number*
Fancy	107 above	16
Special	90 to 107	14
Standard	81 to 96	12
Utility	81 below	10

C. On the Basis of Floral Size

Type	*Floret size (cm)*
Miniature	Below 6.4
Small	6.4 to 8.9
Decorative	8.9 to 11.4
Standard as large	11.4 to 014.0
Giant	above 14.0

D. Basis of Colour

White, Red, Pink, Yellow, Orange, Red Rose, Voilet, Brown etc.

Cultivars

For International Cut Flower Market

Friendship, Hunting Song, Mascagni, Novalux, Peter Pears, Priscilla, Spic and Span, Traderhorn, White prosperity and White Friendship.

Commercial Cultivars

Amarican Beauty, Amsterdam, Apollo, Gold friend, Her Magesty, Morning kiss, Oscar, Pacific and Peter Pears.

Hill Regions

American beauty, Friendship, Gold field, Green wood Packer, White Goddess, Top Brass, Gold Jester and Oscar.

Indian Bred Varieties

Agni Rekha, Aarti, Apsara, Archna, Arun, Basant Bahar, Punjab Morning, Pitamber, Jwala, Mayur and Sher-e-Punjab.

Some Prominent Cultivars (Colour-wise)

- **White:** White friendship, White Prosperity, White Oak and White Wonder.
- **Pink:** Day Dream, Miss America, Pink Triumph, Powder Puff and Rose Suprime.
- **Green:** Green bay, Armstrong, Green brid and Oasis.
- **Rose:** Summer rose, Wine and Roses and Rose Supreme.
- **Red:** Red Majesty, Intrepid, Florida flame and Black Prince.
- **Yellow:** Jester, Jester Gold, Golden harvest and lce gold.
- **Levender:** Anniversary and Elegance.
- **Salmon:** Jenny lee, Dr Magic and Summer garden.
- **Orange:** Peter pear, Tiger flame and Orange Beauty.
- **Voilet:** Tropic sea, China Blue and Blue hawaii.
- **Brown:** Brown beauty and Little tiger.
- **Cream:** Ruffled lotus, Dew drop, Lady Bountiful and Pale moon.

Soil

Sandy loam soil is preferred for the cultivation of gladiolus. The field should be well drained with irrigation facilities. Soil pH between 6.0 and 7.0 is recommended for the cultivation of gladiolus.

Climate

Gladiolus can be grown in a wide range of climates from hills to plain in India but temperature plays a very important role in sprouting, growth and development of the crop. The optimum temperature for growing gladiolus is 19-23°c during the day and 15-18°c in the night. Temperature falling below 6°c may cause frost injury of the plant. It requires full sunlight for proper growth and development.

Propagation

Gladiolus are propagated by corms and cormel. Numerous large and small spawn which are produced around the mature corms or at the end of the short underground shoots from the base of newly-formed corms, are utilized for clonal propagation. In each new growing season, a new corm is produced from the mother corm.

Preparation of Field

The field should be prepared well up to depth of 15-20cm. In Northern plains, after rainy season 4-5 ploughing are desirable. Before planting of corms, the field should be irrigated and then two ploughings are sufficient.

Manures and Fertilizers

It is always better to do green manuring by sanhemp or dhaincha during rainy season in Northern plains. One month before planting, apply FYM 10 kg per square meter. An application of about 50 g bone meal per square meter is often useful. For the good growth and attractive crop, apply 50g Nitrogen and 20g each of phosphorus, potash and Neem cake per square meter. Alternatly 200 to 500 kg/ha nitrogen should be applied in two split doses, while 200 kg/ha each of phosphorus and potash is optimum.

Planting

September to November is the best time for planting of gladiolus to take better crop under Northern plains. The planting distance between rows should be 30-40 cm while the corms should be spaced out 10-20 cm apart and 10-15 cm deep in the soil. According to the size of corm and space for planting 50-75 thousand corms are required for an acre planting, corms should be treated GA_3(100-250ppm) or urea, before planting to Break dormancy. Alternatively, cold stored corms for a period of at least one month is good for planting.

Earthing Up

After about 40-50 days of planting, earthing up should be done to check the lodging of the plants at the time of emergence of spikes due to strong wind.

Irrigation

Light irrigation may be done when ever required to keep the soil moist. Watering should be stopped at the time of repining of corms.

Weeding

Manual weeding is beneficial for weed control. Three to four weeding are good for better crop.

Cutting of Spike/Harvesting

For local market, the flower spikes are best cut when the bottom one or two florets are opening, the remainder of the florets will open successively in the house giving a blooming period of 10-15 days. However, for distance market, the spike are harvested when lower 1-2 florets starts showing colours and still unopened. Gladiolus flower can be harvested 65-120 days after planting, depending upon the cultivar, size of corms and the environment. The spikes may be cut above the four leaves from the base with the sharp knife or scissors after the first floret on the spike has opend. The cutting of the gladiolus spikes is usually done early in the morning and immediately they should be kept in fresh water.

Curing

Curing is one of the essential post harvest operations for successful storage of corms. After lifting and removing the adhering soil, the corms and cormels of each cultivar are kept in trays in a shady but well ventilated place for about a fortnight. For curing, the layers of corms should not exceed three, which may be cured for five weeks at 21°C.

Lifting of Corms

The digging of corms should start when the leaves of the plant become dull coloured or before the plant turns brown. The corms are forked out the ground and made soil free. After lifting corms and cormels should be treated with Bavistin (2 g/ lit. water) and then dried in the shack. The dry, corms are spread out in dry, cool and airy places. When well dried, they should be separated size wise, sprayed against insect-pests and collected in bags or boxes and stored till required for planting next season.

Cleaning, Grading and Storage

After the corms are fully cured, these are cleaned and diseased ones discarded. The old withered corms are taken out and cleaned. Treating the corms with 0.2 per cent Captan 15 days before storage or dusting with 5 per cent Cythione dust and Dithane M-45 protects them from insects, pests and diseases during storage. After cleaning, the corms and cormels are graded in different grade sizes. The corms are stored in perforated trays in a well-ventilated cool and dark room with temperatures are packed in hessian cloth bags before putting them in perforated trays for keeping them in the cold storage. Like corms, the cormels should also be taken out of the cold storage in the first week of October and kept at room temperature for a week before planting them in the ground.

Yield

The flower spike yield in gladiolus vary according to the cultivar, corm size, planting density and management practices. Generally, one plant produces single marketable spike and plantable corm. High plant density with proper management have been found to give higher yield per unit area. Approximate yield of flower spike would be around 2,00,000 per hectare.

Yield of Corm and Cormels

The yield of gladiolus corm and cormels are influenced by cultivar, corm size and other factors. The large cormels produces larger corms with more cormels. 33.1 ton/hectare of corm and cormels when planted at 10×30 cm can be obtained.

Grading

Gladiolus spike are graded into following five grades based on overall quality, length of spike and number of florets in each spike.

Grade	*Spike Length (cm)*	*Minimum Floret Number (Minimum)*
Fancy	above 107	16
Special	96 to 107	14
Standard	81 to 96	12
Utility	81 below	10

Packaging

For local market spikes may be taken submerged in water but for distant markets, these may be carried dry in cardboard and this way, these can easily be retained for 12 hours. Bundles of 50 to 100 spikes are prepared for air lifting in the perforated and light-proof carboard boxes.

Storage

Flower spikes can be stored dry by wrapping in moisture-proof material for two to three weeks at 3 to 4°C, or can be stored wet at 3 to 5°C for 10 days in holding solution containing 4 per cent sucrose plus 250 ppm 8-HQC.

Insect-Pests

Thrips (*Taeniothrips simplex*)

This is a major pest in gladiolus and causes serious damage to the crop. Yellow nymphs and black adults damage leaves and spikes by rasping tissue and sucking the sap. Affected leaves and spikes develop silver coloured streaks which later turn brown, get deformed and dry if the damage is severe. It also attacks corms in storage and infected corms become sticky, shrivel and produce weak plants when planted. It can be controlled by weekly spraying of Rogor 0.2 per cent or Malathion 0.1 per cent.

Aphids

Several species of aphids like *Aphis craccivora, Aphis gossypii* and *Macrosiphum gossypiiattack* gladiolus. They suck the sap from tender shoots and produce flowers of low quality. They also transmit viral diseases. Aphids are slow moving and plump bodied insects. The colour varies with the species and green yellow, pink, brown and black forms occur. Malathion (0.1 per cent) spray checks the infestation.

Cut Worms (*Agrostis segetum*)

Attack different parts of the plants and cut the basal region at the younger stages: Incidence of cut worms is normally observed in the first month of crop. Eggs of moth are seen near the ground level on plant parts. Larvae feed on emerging shoots and cut the plants at the ground level during night. Sometimes they damage underground corms and developing spikes. Spraying of Malathion (0.1 per cent) or Quinolphos (0.05 per cent) at fortnight interval provides protection.

Mealy Bug

Both nymphs and adults attack corms by sucking sap causing shriveling and drying of affected corms in severe cases. Incidence of mealy bugs starts in the field during dry conditions and then it is carried to the storage. Movement of ants on plants is the sign of mealybug infestation. Methyl Parathion 0.04 per cent or Di Methoate 0.04 per cent or Acephate 0.1 per cent at fortnight interval can be applied to effectively control mealy bug.

Leaf Eating Caterpillar (Spodoptera litura)

Egg masses covered with hair are seen on under surface of the leaves. Early instar larvae feed on lower surface of leaves by scraping while mature larvae eat leaves voraciously during night time. Skelitenization of leaves is the main symptom. Spraying of Quinolphos 0.05 per cent or Chlorpyriphos 0.05 per cent or Carboryl 0.1 per cent at 10 days interval controls the incidence effectively. Neem oil 1 per cent or Neem kernel extract 4 per cent checks damage caused by early instar larvae.

Borer (Heliothis armigera)

Larvae feed on leaves and unopened buds. Spraying of Thiodon 0.5-0.8 per cent, or Methyl Parathion 0.05 per cent or Ekalux 0.5-0.8 per cent at fortnight interval can control the pest. Neem kernel extract 4 per cent or Neem oil 1 per cent can also control the pest.

Nematodes

It causes damage to corms and roots of gladioli. Hot water treatment of corms and soil fumigants are remedial measures.

Diseases

Dry Rot or Neck Rot

It causes premature yellowing of leaves. Later leaves turn brown from the tips downward and decay at the basal region. Hot water treatment of cormels and spraying with $CuSO_4$ solution are control measures for the disease.

Storage Rot

It produces redish brown lesions on the corms. Treatment with Benomyl (0.2 per cent) and proper curing of the corms.

Leaf Blight

It produces small, circular yellow lesions with redish centres on the leaves. Three sprays of Dithane M -45, (0.2 per cent) are used for control.

Fusarium Wilt (Fusarium oxysporum f. sp. Gladioli)

Wilt is a major fungal disease in gladiolus. It is soil borne and spread through corms from season to season. High temperature, high level of nitrogen, anaerobic condition and accumulation of carbon dioxide (CO_2) favour the fungus. Roots have brown spots or general rot. Older leaves yellow. Flower size, shape and colour may be abnormal. Flowers may not develop while stalks are curved in an S-shape. Corms rot from the center outward, Oval, sunken spots on the corm surface are brown and may have concentric rings. Do not plant infected corms. Maintain a soil pH of 6.6-7.0 and use nitrate as the nitrogen source when fertilizing. Bavistin (2 per cent) spray and use of resistant varieties are remedial measures for this disease.

Botrytis (*Botrytis gladiolorum*)

Brown spots that may have reddish margins develop on leaves. Spots can be very small to over 1.25 cms in diameter. Similar spots on stems become soft and rot in wet weather. Small, clear spots on petals become brown. A neck rot may occur at the soil line. Small black granules (sclerotia) form on the surface of the infected corms. Do not plant infected corms. Purchase corms that were treated with hot water and fungicides. Apply a fungicide to protect plants. Removal and destruction of the disease plant part and spraying with Manab (0.2 per cent) are controlling measures for this disease.

Curvularia Blight

It is caused by many fungi but the most important are *Curvularia trifolii* and *Curvularia ergrostidis*. Water soaked oval to elongated brownish spots appear on leaves, sheaths and petals and later they coalesce in advanced stage. The patches turn to brown and finally black. Moist and warm conditions spread the blight very fast. It is effectively controlled by Mancozeb 0.2 per cent spray at 10 days interval.

Corm Rot

Corm rot is a common problem caused by group of fungi namely *Fusarium, Curvularia, Stromatinia* and *Pencillium* spp. It causes heavy loss particularly during storage. Anarobic conditions, storage in air tight rooms and more humidity favour the infection. Black, brown, greenish or yellowish mouldy growth on corms is seen during the storage. Under poor air circulation, the corms may rot and emit foul smell. Hot water treatment (38-40°C) containing 2.5g each of Benlate and Captan for 30 minutes has been quite effective. Damage to corms at the time of lifting, improper curing and dampness in storage should be avoided.

11

Jasmine

Jasmine is a genus of shrubs and vines in the olive family Oleaceae. It contains around 200 species native to tropical and warm temperate regions of the Eurasia, Australasia and Oceania. Jasmines are widely cultivated for the characteristic fragrance of their flowers. A number of unrelated plants contain the word "Jasmine" in their common names. Jasmines can be either deciduous or evergreen and can be erect, spreading, or climbing shrubs and vines. Their leaves are borne opposite or alternate. They can be simple, trifoliate or pinnate. The flowers are typically around 2.5 cm in diameter. They are white or yellow in colour, although in rare instances they can be slightly reddish. The flowers are borne in cymose clusters with a minimum of three flowers, though they can also be solitary on the ends of branchlets. Each flower has about four to nine petals, two locules and one to four ovules. They have two stamens with very short filaments. The bracts are linear or ovate. The calyx is bell-shaped. They are usually very fragrant. The fruits of jasmines are berries that turn black when ripe. The basic chromosome number of the genus is 13 and most species are diploid

(2n=26). However, natural polyploidy exists, particularly in *Jasminum sambac* (2n=39), *Jasminum flexile* (2n=52), *Jasminum mesnyi* (2n=39), and *Jasminum angustifolium* (2n=52).

Origin

Jasmines are native to tropical and subtropical regions of Eurasia, Australasia and Oceania, although only one of the 200 species is native to Europe. Their center of diversity is in south Asia and southeast Asia. A number of jasmine species have become naturalized in Mediterranean Europe. For example, the so-called Spanish jasmine (*Jasminum grandiflorum*) was originally from Iran and western south Asia and is now naturalized in the Iberian peninsula. *Jasminum fluminense* and *Jasminum dichotomum* are invasive species in Hawaii and Florida. *Jasminum polyanthum*, also known as White Jasmine, is an invasive weed in Australia.

Taxonomy

Species belonging to genus *Jasminum* are classified under the tribe Jasmineae of the olive family Oleaceae. *Jasminum* is divided into five sections—*Alternifolia, Jasminum, Primulina, Trifoliolata* and *Unifoliolata*. The genus name is derived from the Persian *Yasameen* ("gift from God") through Arabic and Latin.

Uses of Jasmine

Widely cultivated for its flowers, jasmine is enjoyed in the garden, as a house plant, and as cut flowers. The flowers are worn by women in their hair in southern and south east Asia.

Jasmine Tea

Jasmine tea is consumed in China, where it is called jasmine. *Jasminum sambac* flowers are also used to make jasmine tea, which often has a base of green tea or white tea but sometimes an Oolong base is used. Flowers and tea are "mated in machines that control temperature and humidity. It takes four hours or so for the tea to absorb the fragrance and flavour of the jasmine blossoms and for the highest grades, this process may be repeated as many as seven times. It must be refired to prevent spoilage. The spent flowers may or may not be removed from the final product, as the flowers are completely dry and contain no aroma. Giant fans are used to blow away and remove the petals from the denser tea leaves.

Jasmine Syrup

Jasmine syrup, made from jasmine flowersis used as a flavouring agent. It is used for the purpose of perfumes/Drugs/Oils/and much more.

Jasmine Essential Oil

Jasmine is considered an absolute and not an essential oil as the petals of the flower are much too delicate and would be destroyed by the distillation process used in creating essential oils. Other than the processing method. It is essentially the same as an essential oil. Absolute is a technical term used to denote the process of extraction. It is in common use. Its flowers are either extracted by the labour intensive method of enfleurage or through chemical extraction. It is expensive due to the large number of flowers needed to produce a small amount of oil. The flowers have to be gathered at night because the odour of jasmine is more powerful after dark. The flowers are laid out on cotton cloths soaked in olive oil for several days and then extracted leaving the true jasmine essence. Some of the countries producing jasmine essential oil are India, Egypt, China and Morocco.

Jasmine Absolute Used in Perfume and Incense

Many species also yield an absolute, which is used in perfumes and incense. Its chemical constituents include methyl anthranilate, indole, benzyl alcohol, linalool and skatole.

Jasmine as a National Flower

Several countries and states consider jasmines as a national symbol. They are the following:

Hawaii

Jasminum sambac ("*pikake*") is perhaps the most popular of flowers. It is often strung in leis and is the subject of many songs. Indonesia: *Jasminum sambac* is the national flower adopted in 1990. It goes by the name "*melati putih*" and is the most important flower in wedding ceremonies for ethnic Indonesians, especially in the island of Java. Pakistan: *Jasminum officinale* is known as the "*chambeli*" or "*yasmin*", it is the national flower. Philippines: *Jasminum sambac* is the national flower. Adopted in 1935, it is known as "*sampaguita*" in the islands. It is usually strung in garlands which are then used to adorn religious images.

Medicinal Uses

The Jasmine flower forms a vital ingredient of almost all ayurvedic medicines owing to its diverse curing qualities. Specifically it is used to remove intestinal worms. It is considered to be an apt and biological cure for jaundice and other venereal diseases. The flower buds help in treatment of ulcers, vesicles, boils, skin diseases and eye disorders. The leaves extracts against breast tumours. Drinking Jasmine tea regularly helps in curing cancer. Its oil is very effective in calming and relaxing.

Other Uses of Jasmine Flower

The scented flowers are used for making perfumes and incense. The flowers are also flavor Jasmine tea and other herbal or black tea. Its oil is also used in creams, shampoos and soaps. It is considered to be a great skin toner and conditioner. In India Jasmine flowers are stringed together to make garlands. Women in India wear this flower in their hair. Especially in south India, Jasmine flowers are an integral part of their decoration of hair. Some communities even use this flower to cover the face of the bridegroom. People use the Jasmine flowers as religious offerings to the gods like Lord Shiva and Lord Vishnu. Jasmines are the most common flowers in Indian gardens and also important commercially as flowers of some species are valued as loose flower, cut flower and a few yield essential oil for the perfumery industry. Fourty species of genus *Jasminum* is grown all over India having mainly two growing habits (i) Climber and (ii) Shrubs. However, Pot types are also rarely available.

Species and Cultivars

Mainly there are three species commercially cultivated for their loose flowers and for the extraction of essential oil for the perfume industry.

Jasminum grandiflorum

It is known as chameli and Janti. The leaves are shining dark green with 5-7 leaflets. This is a twining shrub with pendulous branches. Flowers are white with a purplish tinge underneath and highly scented.

Jasminum sambac

It is known as Beta, Motia, or Mogra (Double and Single). The highly scented, white flowers are borne in clusters of 3-12. The flower may be single, semi double and perfectly double.

Jasminum auriculatum

It is commonly called as Jooee (Zohi/Juhi) and has smooth foliage with three leaflets. The flowers are white, single or double and are borne in cluster having many flowers.

Others Species

Jasminum abyssinicum, Jasminum adenophyllum, Jasminum angulare, Jasminum angustifolium, Jasminum auriculatum, Jasminum azoricum, Jasminum beesianum, Jasminum dichotomum, Jasminum didymum, Jasminum dispermum, Jasminum elegans, Jasminum elongatum, Jasminum floridum, Jasminum fluminense, Jasminum fruticans L., Jasminum grandiflorum L., Jasminum humile L., Jasminum anceolarium, Jasminum mesnyi, Jasminum multiflorum, Jasminum officinale L., Jasminum polyanthum, Jasminum sinense, Jasminum subhumile, Jasminum subtriplinerve, Jasminum urophyllum.

Cultivars

Jasminum auriculatum

- ☆ **Parimullai:** It is an important cultivar of *Jasminum auriculatum* grown in Tamilnadu, medium round flower bud, white in colour, corolla tube is moderate.Resistant to major insect-pests and diseases. Average flower yield is 8 ton/ha. The duration of flowering is 9-10 months.
- ☆ **Co-1:** The flowers have long corolla tube and this character is helpful in picking of flowers. The average flower yield 8.8 ton/ha.
- ☆ **Co-2:** This cultivar also has long corolla tube and plants are resistant to many diseases, insect-pests mainly to powdery mildew. The average flower yield is 11.10 ton/ha.

- ☆ **Arka Surbhi:** It is a clonal selection from locally grown *Jasminum grandiflorum*. Flower yield is 10 t/ha and concrete recovery is 0.35 per cent. The cultivar has remarkable tolerance for drought condition.

Jasminum grandiflorum

- ☆ **Co-1:** This cultivar is suitable for loose flower for making garlands and for oil extraction purpose. The average flower yield is 10 ton/ha and the recovery of concrete is about 0.29 per cent.
- ☆ **Co-2:** This is a gamma ray induced mutant of CO 1. Buds are white with pink tinge, long and bold. This cultivar is suitable for making garland.The flower yield is 11.68 ton/ha and concrete recovery is 0.29 per cent.
- ☆ ***Jasminum sambac:*** Gundumali- It has round flowers with good fragrance. High yielder. Khoya – Flowers are white and buds are bolder. It has less fragrance. The average flower yield is 8 ton/ha.

Soil and Climatic Requirements

The loamy garden soil is best suited for the cultivation of all species and cultivars with ample water supply. Jasmines grow well between 5-8 soil pH. Water logging conditions damages plants of jasmine. All Jasmines prefer mild tropical climate for proper growth and flowering. *Jasminum officinale* requires 11-27°C temperature for proper growth and flowering. *Jasminum officinale* var.*Grandiflorum* grows best at a temperature of 13°C or less during night. *Jasminum sambac* grows well at a night temperature above 15°C and a day temperature below 20°C. Jasmine is commercially grown in India under open field condition, the ideal requirement for successful cultivation of these plants are mild winter, warm summer, moderate rainfall and sunny days.

Propagation

Generally, all types of Jasmines are propagated by stem cutting and layering. Some of the species like, *Jasminum arborescens* and *Jasminum multiflorum* are propagated from suckers. Initially some rare cultivar may also be multiplied through tissue culture and seeds. In *Jasminum auriculatum* propagation by stem cutting takes longer time as compared to other methods. It takes 78 days to rooting. In general, some hardwood and terminal cuttings are used for propagation with about 70 per cent success. In *Jasminum grandiflorum*- Apical cutting found to be the best for propagation. Ninety four to Ninety eight per cent success was recorded. In *Jasminum sambac* the cuttings treated with the combination of 750 ppm of NAA and 75 ppm lanolin resulted in 90 per cent rooting. TIBA 400 ppm significantly increases per cent rooting in hardwood, semi hardwood and soft wood cuttings in all kinds of Jasmines.

Seeds

To obtain jasmine seeds, either buy them online or from a garden supply store or collect them yourself from a mature plant. Before buying or collecting, do some research on the type of jasmine plant you want. There are over 300 different plants which are known by the jasmine label and all are slightly different. Some ares shrubs, some climb, some have a strong fragrance, some are highly toxic to humans and pets. In the late summer jasmine plants will produce a bean-like seed pod. They can break open at any time and spill seeds everywhere so the best method of collection is gently attaching a sandwich baggie right above the seed pod using a twist tie. This way, when the pod explodes you will catch all of the seeds. When the seed pod begins turning brown you will know it is mature and is about to expel its seeds. If you want to start your seeds inside and move them outside in the spring, start the seeds about 6 weeks before your last hard frost. If you do not want to move your jasmine outside you can start the seeds at any time of year. Soak the jasmine seeds in warm water overnight before planting. To improve success of the seeds germinating you should start them off in seed trays. Use a seed starting mix and cover the seeds lightly with soil. Keep the seeds at count in 0°C with 8 to 10 hours of indirect sunlight a day. If you cannot maintain this temperature place a heating pad set on low underneath the seed trays. Fill a clean spray bottle with filtered or rain water and mist the seeds daily. Never allow them to dry out or become too soggy. Seeds can be slow to germinate and may take as long as one month.

Cuttings

Another simple way of propagating jasmine is by growing it from cuttings. This is actually much easier than many people believe. Start by taking a cutting from a plant when there are no flowers on it. Ideally, this should be done in late summer to early fall. Using sharp scissors, take a 5 or 7.5 cms. section of jasmine and snip it off the plant right below a leaf. Make the cut at a 30 degree angle. If you're trying to grow jasmine by taking cuttings, you need to take more than you need. Not all of these cuttings will take and it's always better to have more than you need. Remove the bottom leaves to leaf nodes. Then dip the cut end of the cutting into rooting powder. Press the cutting into a moist sanitized medium like sterile potting soil, vermiculite, perlite or peat ideally a combination of these. Never use garden soil, which may be contaminated with bacteria or harmful microbes. Plant the cutting at a depth of 2.5 cm. Place your container in a room that gets good light but no direct sun. Keep the cuttings at 21.2°C, using a heating pad set to low underneath the pot if this temperature is not possible. Fill a clean spray bottle with filtered or rain water and mist the cuttings daily to keep them moist and keep the humidity up. Alternatively, cut some ventilation holes in a ziplock baggie and place the baggie overtop of the cutting to maintain a high level of humidity.

Layering

In this method, a branch of the plant is pulled towards the ground and a part of it is covered with moist soil leaving the tip of the branch exposed above the ground. After some time, new roots develop from the part of the branch buried in the soil. The branch is then cut off from the parent plant. The part of the branch which has developed roots grows to become a new plant. Chameli is propagated by the layering method. One left side branch and one right side branch of the parent jasmine plant have been buried in moist soil. The parts of branches which are buried in soil grow their own roots. When this happens, the branches of the parent plant connecting the newly formed plants are cut off so that the newly formed plants may grow on their own and develop into mature plants. The natural layering occurs because these plants form runners. Wherever the ends of such runners touch the ground, new plants are formed at those places.

Suckers

Loosen the soil around the sucker with a fork and then carefully expose and lift the sucker and associated roots, being careful not to disturb the parent plant. Using sharp secateurs or a knife, sever the sucker, making sure that it has fibrous roots on the detached portion. Replace and firm the soil around the parent plant. Trim the sucker by removing the main root or stolon up to the fibrous roots. Reduce long, leafy shoots by about half to limit drying out of the sucker after planting and to promote bushy re-growth. Plant the sucker into fertile, free-draining soil enriched with organic matter such as well-rotted garden compost, manure or leafmould. If potting up the sucker, multipurpose compost is fine to use. Water well after firming the soil around the roots. The roots are usually insufficient to sustain the plant without careful watering for the first season.

Tissue Culture

A rapid and improved procedure of micropropagation of jasmine (*Jasminum azoricum* L.) was studied. Different types of tissue explants were cultured on MS salt mix containing different concentrations of naphthalene acetic acid (NAA), benzyl-aminopurine (BAP) and kinetin. The explants culture was conducted at different times in order to assess time effects on bud proliferation, elongation and root formation and to obtain proper clones after acclimatization. The results showed that kinetin (0.5 mg/L.) on the MS medium gave the best result with regard to bud proliferation and multiplication. However, addition of NAA of different concentrations was inhibitory. The response of base and intermediate cuttings was better than terminal ones. The best time for bud proliferation in primary explants culture on MS medium supplemented with 0.5 mg/L kinetin was February and for bud multiplication February and March. Using 1.5 mg/L IBA in *in vitro* rooting medium, gave 97; rooted plantlets, while *ex vitro* rooting of plantlets on agricultural media after soaking their bases in a solution containing 20 mg/L IBA increased rooting percentage up to 99 per cent. In addition, the *ex vitro* rooting resulted in successful hardening of 94 per cent of the plantlets and reduced time, costs and efforts required for rooting and hardening

Preparation of Field

While selecting the site for the cultivation of Jasmine, it should be kept in mind that the area should be open having sunny situations without any interference of trees or buildings and the site should have proper drainage and irrigation facilities. Before planting, pits of the size of 45x45x45cm are prepared at least one month prior to planting and normally, these pits are dug in the month of April. They should be exposed to sunlight at least for one month. Few days before planting these pits are filled with mixture of 2 parts FYM+ 1 part fresh soil + 1 part sand and 50-100 g of any insecticide powder like folidol dust. After filling, pit should be irrigated for settling down of soil and one healthy plant should be plant in the centre of the pit. *Jasminum* plant may be planted of 3-4 years interval. At the time of planting proper distance, proper soil treatment is necessary. In *Jasminum auriculatum*, a spacing of 1.80×1.8 metres is kept and for proper plant growth and flowering, In case of *Jasminum grandiflorum*, a spacing of 2m x1.5m accommodating 3333 plant/ha recorded maximum flower yield of 13.55 ton/ha. A spacing of 1.2×1.2 metres produce quality flower and optimum yield is needed in *Jasminum sambac*.

Nutritional Requirement

For newly planted plants, 100g N, 150g P_2O_5 and 100 g K_2O are given. If soil is poor in nutrition, then 50 per cent amount of all these fertilizers is given at a time of filling of mixture. Among the micronutrients, 10kg/ha Zn, 40 kg/ha Mg and 2 kg/ha Mo are given. Generally, these micronutrients are given as foliar application at different stages of the growth at regular intervals. The requirements of NPK among the different species of Jasmine are as follows:

Species	*N/ha*	*P/ha*	*K/ha*
Jasminum auriculatum	60	120	120
Jasminum grandiflorum	120	240	120
Jasminum sambac	60	120	120

Irrigation

A sufficient amount of moisture should be maintained for proper growth and flowering throughout the growing period. During blooming, the water should be applied twice a week if there is no rain and once in a weak during rest of months.

Pruning

Pruning is an important cultural practice in Jasmine which encourages growth of new healthy shoots and influences the flower yield. In this regard, time and intensity of pruning are important. Jasmines are pruned by heading back all the past season's shoot at a height of 50 to 90 cm from ground depending upon the species and all the leaves from the remaining parts are stripped off. While pruning, all unhealthy shoots are to be removed retaining only 9 to 10 shoot per plant. The different *Jasminum* species have different pruning requirements.

Jasminum grandiflorum: In this case, December pruning is beneficial as maximum yield is obtained if the plants are pruned in this month and first picking could be done in March.

Jasminum auriculatum: They should be pruned during January to March to have flowers for longer duration.

Jasminum sambac: They should be pruned during January and also can be pruned after normal flowering season which stimulates another flush.

Chemical Defoliants

Pruning is essential but time consuming operation in commercial cultivation of Jasmine. In addition to the labour cost involved, manual pruning can result in injury to the plant and is a potential source of outbreak of diseases. Chemical defoliants reduce dominance and thus encourage the simultaneous growth of lateral shoots. In *Jasminum grandiflorum*, spray of pentachlorophenol at 3000 ppm causes a significant defoliation and improves flower yield over manual pruning.

Harvesting

Jasminum plants give economic yield after third year and goes up to 12-15 years if the plants are properly maintained. The right stage of harvesting of flower is determined on the basis of purpose. When flowers are fully developed and opened flower buds are harvested early in the morning for oil extraction purpose. For other purposes like veni and gajra, newly opened buds are harvested. The buds are harvested for other decoration purposes.

The average flower and concrete yield recorded at IIHR varies with species and varieties:

Yield

Table 11.1: The Yield of Flower, Concrete Recovery and Concrete of various Species of Jasminum

Species	*Flower Yield (Kg/ha)*	*Concrete Recovery (Per cent)*	*Concrete Yield (Kg/ha)*
Jasminum auriculatum	4636 to 9022 kg/ha	0.28 To 0.36	13.44-28.44kg/ha
Jasminum grandiflorum	4239 to 10144kg/ha	0.25 to 0.32	13.85- 29.42 kg/ha
Jasminum sambac	2063 to 8129 kg/ha	0.14 to 0.19	11.18-15.44 kg/h

Post-harvest Treatment

Due to short span of life of harvested flowers, the growers often face the problem of storage in the peak season of flowering. Fully developed flower buds if soaked in any of the chemical solutions, namely, sucrose(1 per cent), boric acid (0.5 per cent), copper sulphate (0.1 per cent) aluminum sulphate (0.1 per cent) or silver nitrate (0.1 per cent) remain fresh for two days without losing the fragrance. Cold treatment also enhances the longevity of flowers. Packaging of Jasmine flower in polyethylene bags (200 gauges) without ventilation is beneficial in maintaining higher quality and extending the shelf life of flowers. Storing the packed flowers in cool chamber (8°C) further enhances the flower quality and shelf life. In order to induce colour, the flowers are soaked in edible colour solutions, the process is called tinting.

Grading

There is no standard grade available for Jasmine. The flowers may, however, be graded according to the corolla tube length, buds size and freshness.

Packing

Wholesalers pack flowers in bamboo baskets. They are repacked so as to maintain some moisture and air circulation in the baskets. Water is sprinkled on the newspapers covering the inside of the basket. The top is covered with paper again and closed with bamboo basket cover or sunny sac which is stitched to the edges. No time should be lost in transporting flower after harvest. The flowers are sent to the destinations through various means of transportation including reefer vans for long distance transportation.

Plant Protection

Pest Management

Bud Borer

Bores into the immature flower buds and feeds on floral structure. In severe cases webbing of bud is also noticed. Spraying of metasytox (0.2 per cent) has been found effective in combating the insects.

Red Spider Mites

Incidence of this pest is high during warm and dry weather. They feed on under surface of the leaves which becomes yellow and finally drop off.Spraying of common sulphur (2 g/l) is required for control.

Nematodes

Root knot nematode, *Meloidogyne incognita* causes swelling of roots and the rootlets become enlarged leading to conspicuous pale yellowing of leaves and die back. Phormese applied @4g/plant during the month of May and September followed by pruning in December incorporating 20kg FYM/plant increased the yield of Jasmine by reducing the rootknot nematode population by 70 per cent.

Disease Management

Leaf Blight

Cercospora jasminicola and *Altenaria Jasmini* cause leaf blight disease. Reddish brown circular spots produced on the upper surface of leaves. Affected leaf margins show inward curling and become hard and brittle. Disease appears in the month of May/June and may cause 50 per cent loss in yield. The spray of 0.4 per cent Benlate, 0.2 per cent Dithane, M-45, 0.1 per cent Bavistein and 1.0 per cent Bordeaux mixture have been found equally effective to control leaf blight.

Wilt

It is caused by *Fusarium solani*. Yellowing of the lower leaves is the early symptoms of this disease. Yellowing gradually spreads upwards, finally resulting in death of the plant. Drenching the soil around the plants with 1 per cent Bordeaux mixture controls this disease.

12

Lilium

Lilium belongs to family Lilleaceae and are believed to have emerged 55 million years ago and has been in culture for the last 3,000 years. Plot of lilies made its entry and was very popular in Egypt's 18th dynasty. Many bulbous plants called lilies. Many of them, however, belongs to other genera. The here mentioned all plants belongs to the genus Lilium. All modern lily hybrids derive from approximately 100 lily species, of which approximately a dozen tribes from Europe, Approximately two dozen from North America, approximately 40 from China and the remaining 25 come from Japan, Nepal, Myanmar and Korea. The botanic name Lilium is the Latin form and is a Linnaean name. The term "lily" has in the past been applied to numerous flowering plants, often with only superficial resemblance to the true lily including water lily, fire lily, lily of the Nile, calla lily, trout lily, kaffir lily, cobra lily, lily of the valley, daylily, ginger lily, Amazon lily, leek lily, Peruvian lily and others. Lilium is a genus of herbaceous flowering plants growing from bulbs all with large prominent flowers. Many other plants have "lily" in their common name but are not related to true lilies.

Origin and Distribution

Many species are widely grown in the garden in temperate and sub-tropical regions. Most species are native to the temperate northern hemisphere, though their range extends into the northern subtropics. The range of lilies in the Old World extends across much of Europe, across most of Asia to Japan, south to India and east to Indochina and the Philippines. In the New World they extend from southern Canada through much of the United State.

Climate

The optimum temperature during growth should be maintained between 10-25°C. Higher temperature above 35°C will produce short plants with less number of flower buds/Spike. A low initial temperature 9-13°C is recommended especially with a view to good rooting. Lilies thrive best in sunny conditions. When the light is restricted they lean towards the light. Very few lilies can survive where the shade is dense.

Botanical Description

Chromosome no 2n= 24, it is a perennial, erect, leafy stemmed herbs with underground scaly bulbs. Flowers are pendulous, inclined, horizontal, or erect with six separate segments, which are scarcely differentiated as between sepal like and petal like organs, each bearing a nectar groove or furrow at the base. Stamens are six in number, hygynous or slightly adherent to perianth mostly shorter than the segments. Anthers are versatile, filaments slender, pistil one with long style and three lobed stigma, fruit dry, loculisidal and many sided capsule.

Uses

As a cut flower in decoration purpose, may be planted in mixed border. They are excellent in flower arrangement and bouquet making. They may also be grown as potted plants. Numerous ornamental hybrids have been developed. They can be used in herbaceous borders, woodland and shrub plantings and as patio plants. Some lilies, especially *Lilium longiflorum*, form important cut flower crops.

Species

Asiatic Hybrids

These lilies are hardy and very easy to grow. They come in all shades and colour combinations. These hybrids multiply rapidly and bloom over a long season. The flowers can be up-facing, side-facing or down-facing, vary in height and flower mid season.

Martagon Hybrids

These dainty flowered ones are fast becoming popular with as many a 50 flowers on tall erect stems with whorled leaves. These lilies often take a year to settle in and are known to sulk the first growing season, sometimes not showing any presence until the next growing season. Once established in the border or garden they can be left alone for many years. They are the only lilies that prefer dappled shade.

Orientals, Trumpet and Aurelians

These varieties are the most exotic and showy among all lilies. These varieties multiply more slowly, it takes a little more effort on your part of amending the soil and providing heavier winter protection of straw, leaves and peat moss. Protecting them from the first frosts, by covering them with a cardboard box will allow the bulbs to mature more fully for the following year. These large beautiful scented flowers that bloom late summer will be well worth the extra effort needed.

Orienpet - Longipet - L.O. Hybrids

These lilies are breakthroughs in the lily hybridizing world giving improved hybrid vigor and large showy flower. They are proving more hardiness then their parents, the orientals, trumpets and longiflorums. However, these should be mulch to aid in over wintering in cold climates

Cultivars

- ☆ **Lilium Asiatic:** Holeaw America (Red), Tressor (Orange), Black Out (Red), Brunello (Orange), Heraclion (Orange), Montenegro (Red), Navona (White).
- ☆ **Lilium L.A hybrids:** Ceb Dazzle (Yellow), Brindisi (Pink), Nova Scotia (White), Algarve (Pink), Cosenza (White), Courier (White), Ercolano (White), Fangio (Red), Sulpice (Hot Pink), Royal Trinity (Peach), Original Love (Orange), Pavia (Yellow), Red Alert (Red), Orange Tycoon (Orange), Couplet (Pink).
- ☆ **Lilium oriental:** Acapulco (Pink), Conca D'or (Yellow), Siberia (White), Stargazer (Purple), La-Mancha (Pink), Laguna (Pink), Aktiva (Pink), Santander (White), Starfighter (Hot Pink), Tiber (Pink), Cobra (Red), Sorbone (Pink), Alma Ata (White), Red Reflex (Red), Shocking (Bicolour).

Soil Preparation

Lilies prefer slightly acidic, humus-rich soil well drained soil.A soil pH of around 6.5 is good for growth. If drainage is poor, then planting is done in raised beds. Turn over the soil to a depth of 30-45 cms. Work humus and fertilizer into the soil but avoid the use of fresh manure or other fertilizers high in nitrogen which encourages rot problems. Add lime which causes the flocculation of soil particles and allows improved air and water management. Lilies require direct sunlight for part to all of the day. A medium sandy loam soil with a reasonable amount of humus is ideal. Peat moss can also be added. Heavy soils can be lightened with course sand and peat moss. If using manure make sure it is well rotted and use as top dress only. Otherwise it can cause damage to the bulbs by lowering their disease resistance.

Propagation

lilies are propagated through firm scaly bulb that are available round the year. But Indian conditions are deprive of lilium bulbs and hence bulbs are exported from other countries like Holland. As a thumb rule the smaller the bulbs resulting lesser the spike length and number of flowers per stem. For Asiatic hybrid liliums bulbs of 12/14 or 14/16 cm size (measurement of circumference in cm) should be used. For Oriental Hybrid lilies" the bulb size should be 16/18 cm size or more depends on variety. From the moment of arrival of the bulbs, keep them moist and cool at all times. Disinfection of the bulbs before planting is necessary with Bavistin 2 per cent to check bulb rot. For lilium, as they usually arrive in the sprouting stage, it is very important to plant the bulbs immediately upon arrival.

Planting

Lilies are usually planted as bulbs in the dormant season. They are best planted in a south-facing (northern hemisphere), slightly sloping aspect, in sun or part shade, at a depth 2½ times the height of the bulb (except *Lilium candidum* which should be planted at the surface). Most species bloom in July or August (northern hemisphere). The flowering periods of certain lily species begin in late spring, while others bloom in late summer or early autumn. They have contractile roots which pull the plant down to the correct depth, therefore it is better to plant them too shallowly than too deep.Most lilies should be planted to a depth of 15 cm. and 30-45 cm. apart. Trumpets and orientals should be planted to a depth of 20 cm. for extra winter protection. Place your lily bulb with its roots down and scale points up. A little bone meal may be added at this time. Cover with your soil mixture. Pack the soil in well around your bulb. It is important to thoroughly water your bulbs in after planting so the soil settles around the bulb to prevent any air pockets

Planting Time

Most lilies do best when planted in early fall(autumn planting-octuber). But *Lilium candidum* must be replanted in late July or August a few weeks after it flowered.

Container Growing

Bulbs may be planted singly in a 15 cm. pot or in groups of 3 in a larger pot, allow for 10-15 cms. of soil above the bulb. Use a commercial potting soil mixed with 1/3 peat and 1/3 perlite or sand to allow for good drainage. A regular application of fertilizer as well as repotting every year with fresh soil is recommended. Because lilies are never completely dormant, extra care for winter must be taken. Place pots in a root cellar or remove bulbs and store in plastic bags with peat moss in a fridge. Bulbs should not get below freezing.

Fertilizers

The kind and amount of fertilizer depend on the soil fertility. A complete well balanced fertilizer with NPK (15 -15 -15) or 20-2 is more suitable. Apply garden lime if the soil is too acidic. Excessive fertilizer is too harmful to lilies. Fertilizer is best applied when lily shoots are at the spear stage just before the leaves unfurl. An old practice is to scrape away a couple of centimeters of soil just as the lilies emerge and replace it with a mulch of well decayed FYM. Top dressing of Calcium ammonium nitrate (CAN) @ 1 kg./100 m2 done at flower bud development stage. After one month another dose of CAN be given per 100 m2. To avoid leaf burn, it is recommended to irrigate the crop with clean water later on. Always irrigate after fertilizer application.

Replanting

Asiatic cultivars should be lifted and replanted on a regular basis. Most of them form many bulblets among the underground parts of the stem which results in crowding and subsequently flowering decreases. Lifting and division should be done about every third year with stronger Asiatic cultivars. When the lilies are not growing strongly then move to a new location and fresh soil which will help in regaining lost vigor. This planting is best done 3-4 weeks after flowering. Check the bulbs for diseases (fusarium) and discard severely infected bulbs. Treat the bulbs with fungicides.

Irrigation

Lilies do not prefer overhead water. This will cause Botrytis and other fungal diseases. Frequent irrigation is necessary for proper growth and development of plant. The lily bed should never be too dry or too wet. Watering can be done according to the intensity of overhead sun.

Mulching

Material used for mulching depends on the locally available material like rice husk, saw dust, well rotten cow manure, leaf mold, sphagnum peat and mushroom compost.

Weed Control

For lilies manual weeding can be done if the area is small. But for larger grounds, where manual weeding is not possible, use Roundup for control weeds. Apply Roundup before the emergence of flower.

Harvesting

Flowers can be harvested as soon as the first lower two buds start showing colour. For domestic market the optimum stage of harvesting is when the first bud is fully bulged and coloured. In general Asiatic lilies take 3 months while Oriental lilies take 5 months to start harvesting from the time of planting. When harvesting the flowers, cut the stems approximately 15-20 cm above the soil leaving a few healthy leaves on the mother plant. Always harvest the flowers in the early morning. As soon as the flowers are cut, take them to a shady place like a packing shed and put the flowers in bucket of water.

Packing and Storage

Lilium should be supplied in bunches of 10 stems. Lilium should be wrapped in a sleeve per bunch. Immediately after bunching, the cut flowers should be placed in cold water in cold storage room at 2°C to 3°C. Add 2 per cent sucrose and 100ppm GA3 as a preservative agent to water to improve vase life of flower. When dispatching lily flowers use only perforated CFB-boxes to maintain a proper temperature during transport. If supplied in a cut flower container, it is advisable to pre-treat Lilium and supply in water with a preliminary treatment that contains the active ingredient Sodium dichloroisocyanurate. Dosage: 1 chlorine tablet per 3 litres.

Transportation

Lilies should be shipped in perforated boxes. The perforations are necessary to prevent the accumulation of excessive concentrations of ethylene, a hormone produced by the opened flowers themselves. To prevent the premature forcing of flowers and the development of fungi, make sure when packaging that the product goes into the box dry. Low transport temperatures (preferably cooled to between 1 and 2°C) are necessary for lilies to prevent flower bud dropping as well as the adverse effects of ethylene. For long distance transport or export, it would be highly advisable to pre-cool the boxes before they are dispatched. Upon arrival at the wholesaler and/or retailer, the lilies should again be trimmed, placed in clean water and stored at 1 to 5°C.

Handling of Bulb

For harvesting of bulb reduces the frequency of irrigation water. Maintain soil moisture level in such a way that bulb scales should not dry out. Excessive moisture may lead to rotting of bulbs. Allow bulbs to remain in the beds for 4 to 5 weeks (above ground stem portion should dry out and can be pulled out from bulb easily).After 5-6 weeks remove the bulbs from soil along with dried stem. Remove dried stem carefully without damaging the bulb. Wash bulbs with clean water and treat with 2 per cent Bavistin solution for 10 minutes. Remove the bulb from solution and air dry in shade. Immediately after air drying pack the bulbs in plastic crates with moist coco peat wrapped with perforated plastic sleeves. Coco peat used for packing must be sterilized. Keep the crates in cold storage at 2°C for 2 weeks and then at -10C for 6 weeks. Keep crates open for one day in cold storage and then close with plastic sleeves.

Insect-Pests and Diseases

Insects

Lily Beetle

Long prevalent in Europe, the lily beetle (*Lilioceris lilii*) has been reported in eastern North America. The larvae and adult beetles feed on the leaves of lilies. The larva is a yellow grub with a dark head, covering itself in dark, slimy excrement. The adult is up to 8 millimeters long and bright scarlet with black legs and antennae. Both life stages have voracious appetites and soon devour entire plants. The eggs are laid on the underside of the foliage. The following controls are effective: Spray plants with contact and systemic insecticides, both are effective. Drench soil with a soil insecticide to kill the mature larvae that live just under the soil surface in winter. Also, avoid transporting infested soil to other sites. Catch adult beetles between the fingers and smash them. The lily beetle has only appeared in a few places on this continent and with care, it should be possible to prevent any lasting infestation.

Aphids, Capsids and Leafhoppers

Several of the commoner Aphids (Greenfly/blackfly) may attack lilies. Not all species will attack these plants. These insects can travel considerable distances by flying or being swept in the wind. Most of these insects can live of quite a range of host plants and so are often quite widespread. They suck the sap of host plants and can cause distortion of the growing plant especially while the plant is actively growing (Spring). In addition to possible physical damage caused, sap sucking insects can also transmit viruses from plants they were previously feeding on. It is fortunate that many of the most damaging viruses are specific to Lilies or related plants and therefore, if Lilies are grown in separate groups it will reduce the risk of spread. Perhaps the most likely source of serious virus infection is from other Lilies or Tulips. The most commonly transmitted non native virus is Tulip Breaking Virus. It is not a native virus and has a very restricted host range and therefore, is most unlikely to get a toe hold in the wild. Its chief source is from Tulips brought in from tulip growing areas. Some infected tulips do not show symptoms but others do. All Rembrandt Tulips or other "broken" or striped tulips are probably infected.

Diseases

Botrytis

It is a fungus disease, which affects the leaves of lilies, caused by excessive moisture and warm temperatures. The first signs can be white spots on the leaves. In severe cases the whole leaf and stem can become infected and the whole plant decay and collapse. Injury to plants, like frost or hail will make it easier for Botrytis spores to enter the leaf; spraying is strongly advised very soon after injury. The disease is not carried by the bulb so it will not affect flowering the following year. In early stages of infection, if possible remove noticeable spotted leaves. Spraying is highly advisable and is only effective when foliage is dry. A copper spray can be used or natural remedies such as a baking soda mixture (1/4 tea spoonful per 0.25 litre of water) sprayed weekly on the foliage during wet periods can also be used. Good air circulation will help prevent a outbreak. Planting lilies some distance apart will also control infection. In the fall clean up and burn dead stems and leaves.

Basal Rot

This fungus invades the bulb through the roots and basal plate. The symptoms on the growing plants are usually premature streaky yellowing of the foliage. The disease can become present in warm moist soils. As a preventive, avoid over-watering during warm summer months and provide good drainage. As for infected bulbs, one can remove the infected scales, dip the bulbs in a fungicide solution of Benlate which is readily available to home growers.

Blue Mold

Because lilies have a high sugar content, bruising or mechanical injury can cause a penicillium mold to form on the injured part of the bulb. This is harmless to the bulb and can be carefully removed. The bulb can be dusted with a fungicide powder and planted as usual.

Virus Diseases

Lily viruses are transmitted largely by aphids. Visible evidence of virus is irregular mottling and flecking of the leaves. Reduction in plant size and height. Distorted, twisted growth. Colour-breaking in the flowers and leaves. Brown ring patterns on bulb scales. A few tips to help control viruses are: Destroy clumps of lilies that show severe infection, insuring that all bulbs and scales are discarded. Remove plants showing infection early in the season. Avoid planting lilies next to other host plants like tulips or *Lilium tigrinum* (Tiger Lily). Control aphid infections with the use of insecticides.

Lily Thrips (*Liothrips vaneeckei*)

This thrips spends most of its life below ground and has no wings therefore, it is incapable of spreading to new areas without mans help. Only the bulb is effected the pest feeds at the base of the scales causing brown spots. Most of the damage is probably due to secondary infections. Plants may lack vigour, look sick or quickly collapse. There is no sign of pest above the soil. In the soil or attached to the plant very small white insects can be seen, moving fairly slowly. With a hand lens or with very good eyesight on closer inspection, these insects are elongated in shape with three pairs of legs, and look like typical thrips but without wings. The adults are generally more difficult to see but appear a shiny silver and are faster moving, again they look like thrips but have no wings. It is not known how widespread they are but it is likely to become more common with open borders between countries. Although they effect lilies, it is possible they have a wide host range if neccessary, they possibly can effect most "bulbous plants". It is possible these insects have some natural predators. Dipping the bulbs or drenching the plants in insecticides may give some control. A number of insecticides in the past have been used quite successfully but many are now not available. It is possible pyrethroid insecticides will give some control or any insecticide which is effective against thrips, leafhoppers or aphids.

Vine Weevil, Millipedes, Wireworms, Leatherjackets, Chafer Bugs

Millipedes, Wireworms (orange/yellow grubs) and leatherjackets can do damage to the roots and bulbs of lilies and cause secondary infections. They tend to be localised and because they are not usually present in large enough numbers in a given volume of soil bulbs are not usually killed. Symptoms roots are eaten bulbs are holed. There are a whole host of natural predators already mentioned. Wasps as well have some effect particularly on leather jackets. Cockchafer grubs will attack bulbs but usually they are just a nuisance. Vine weevil has been around for some time, however, it rarely attacked lilies until a few years ago. In its current form, some circles it is believed to be a man introduced alien probably from Europe. Since the Garden Centre boom of potted plants and the regular import of large quantities of potted plants with compost from Europe their has been a sudden change. This pest has miraculously developed a taste for almost anything put in front of it including lilies. Even so their does appear to be some differences, damage does appear to be more frequent on container grown lilies in a peat based compost, slightly less on containers with soil based compost and least frequent in the ground. This may be due to natural predators.

The larvae which probably do the most damage live in the ground and eat plant roots and open the plant to secondary infections. The adult weevils eat leaves. smaller plants up to the size of lilies can have so many roots eaten that the plant dies of drought. Recognising the pest signs of its activity are the plant looks unhealthy and lacks vigour and may even wilt suddenly. Leaves of the lily or nearby plants may show scalloped edges as a result of the adult feeding. The adult is a small blackish weevil with a rounded abdomen and pointed head because of its markings the beetle appear matt grey rather than black. It has no wings and is not a great walker, so spread to new areas is only with mans help. In recent years most plant producers have taken this pest very seriously and most growers now as routine treat all their compost with either a special insecticide or a biological pest control organism. The larvae live in the soil and are creamy white with a brown head. The insect is nearly always in a curved position. They breed possibly once or twice a year in this country. A weird thing about this pest is that all of them are female and genetically identical. They are to lay their eggs without fertilisation and do not need males. Although this has certain advantages for slightly quicker reproduction in the short term it does rather leave them open to attack and it may be within a few years a natural predator or a man made one might suddenly wipe the whole species out. Control: These pests do have quite a fair number of British natives gunning for them. Thrushes, Blackbirds, Robins and most of the Crow family are particularly fond of the larvae if they can find them. Ground beetles and nematodes take their toll. In the garden/field it is possible to get adequate control except close to cyclamen. Parasitic nematodes can be bought which will attack the larvae and can give good control for up to one season, if applied as directed. There are also chemicals either mixed into or drenched onto the soil or compost, which can give good control in some cases into the second year from one treatment. Suscon Green, Imidacloprid. Other insecticides will kill vine weevil but it is difficult to get the insecticide close to the larva in the soil, so a percentage will survive. Used as a soil drench Malathion, BHC, Parathion. The adult weevil can be killed by spraying with almost any contact insecticide but it is difficult to kill all the adults. Hand picking of adults is usually not 100 per cent effective either.

Sciarid Flies

These are normally only pests under glass they probably only do minor damage on adult plants, most probably being caused by possible secondary infections. On seedlings and other small plants they are more likely to cause problems. This was followed by the belief that they would attack growing plants only if the compost was too dry. The main attack appears to be on the root hairs. They breed very fast and soon form clouds of flies. They are very small flies with black bodies which sit on the compost till disturbed. The young are small white maggots with dark heads. Malathion, and systemic insecticides are unlikely to give control as a drench. One has tried permethrin, bifenthrin, rotenone (derris), pirimcarb, malathion and metasystox as a spray against the adults with no apparent effect. One has used a vegetable oil spray based on oil seed rape and this gives good control but needs repeated treatment. One has used nematodes to control sciarid but have had no useful effect with repeated treatments.

13

Marigold

Cultivation tends to be located close to big cities like Mumbai, Pune, Bangalore, Mysore, Chennai, Calcutta and Delhi. The estimated area on which flowers are grown in India is about 1,10,000 hectares. Major growing states are Karnataka, Tamil Nadu, West Bengal, Andhra Pradesh and Maharashtra. Traditional flowers including marigolds, occupy nearly two thirds of this area. In northern India, small scale farmers are growing marigold and other flower crops for garlands and decoration. A native of Mexico, marigold has been grown in gardens throughout the world for hundreds of years. Today, they are one of the most popular bedding plants in the United States. Marigolds are easy to grow, bloom reliably all summer and have few insects and disease problems. The marigold's only shortcoming is its pungent aroma. There are numerous marigold cultivars available to home gardeners. Many of the commonly grown marigolds are cultivars of African and French marigolds. Less known are the triploid hybrids and the signet marigolds.

The African marigolds (*Tagetes erecta*) have large, double, yellow to orange flowers from midsummer to frost. Flowers may measure up to 12.7 cms across. Plant height varies from 25.4 to 90 cms. African marigolds are excellent bedding plants. Tall cultivars can be used as background plantings. African marigolds for Iowa include cultivars in the Inca and Perfection series. African marigolds are also referred to as American marigolds.

The French marigolds (*Tagetes patula*) are smaller, bushier plants with flowers up to 5.1 cms across. Flower colours are yellow, orange and mahogany red. Many cultivars have bicoloured flowers. Flower heads may be single or double. Plant height ranges from 15.2 to 45.2 cms. The French marigolds have a longer blooming season than the African marigolds. They generally bloom from spring until frost. French marigolds also hold up better in rainy weather. French marigolds are ideal for edging flower beds and in mass plantings. They also do well in containers and window boxes. Queen Sophia and Golden Gate are excellent French marigold cultivars. Cultivars Early Spice, Hero, Janie and Safari series also perform well in Iowa.

The triploid hybrids are crosses between the tall, vigorous African marigolds and the compact, free-flowering French marigolds. Triploid hybrid marigolds are unable to set seed. As a result, plants bloom repeatedly through the summer, even in hot weather. One problem with the triploids is their low seed germination rate. Average germination is around 50 per cent. Since the triploid hybrids are unable to produce viable seed, they also know as mule marigolds. Signet marigolds (*Tagetes tenuifolia*) are quite different from most marigolds. Signet marigold plants are bushy with fine, lacy foliage. The small, single flowers literally cover the plants in summer. Flower colours range from yellow to orange. They are also edible. The flowers of signet marigolds have a spicy tarragon flavour. The foliage has a pleasant lemon fragrance. Signet marigolds are excellent plants for edging beds and in window boxes. The cultivars Golden Gem and Lemon Gem do well in Iowa. There are basically three planting options available to home gardeners when planting marigolds. Marigold seed can be sown directly outdoors when the danger of frost is past or started indoors 6 weeks prior to the last frost date. Marigolds are also available as bedding plants at garden centers. Planting site requirements for marigolds are full sun and a well-drained soil. Plant spacing varies from 15.2 to 22.8 cms for the French marigolds and up to 45.2 cms for the taller African marigold cultivars. Summer care of marigolds is simple. Water occasionally during dry weather and pinch off faded flowers to encourage additional bloom. Tall African marigolds may require staking to prevent the plants from falling over or lodging during storms. While marigolds are seldom bothered by insects and diseases, they are not problem free. Spider mites can devastate marigolds in hot, dry weather. Grasshoppers can also cause considerable damage. Aster yellows is an occasionally disease problem. In a related matter some gardeners plant marigolds in their vegetable gardens to repel harmful insects. While the marigolds are an attractive addition to the garden, research studies have concluded they aren't effective in reducing insect damage on vegetable crops.

Climate

Both African French marigold are hardy in nature and grow well under tropical and sub-tropical conditions almost throughout the year. It has an added advantage that it is not season bound like other annual flower crops. Marigold needs plenty of sunshine and is, therefore, grown in open sunny situation. The performance in shady location will be very poor. The flowers of marigold can be obtained throughout the year by taking crops in succession in order to ensure their regular supply to the market.

Soil

Sandy loam soil with a pH range of 7.0 to 7.5 having good aeration and drainage is ideal for growing of marigold. The land has to be ploughed well unto a depth of 30 cm nearly a fortnight prior to transplanting of seedlings. Well rotten farmyard manure at the rate of 30 tonnes per hectare should be incorporated well in the soil at the time of soil preparation.

Species

Tagetes patula, Tagetes arenicola, Tagetes erecta, Tagetes argentina Cabrera, Tagetes biflora Cabrera, Tagetes campanulata Griseb, Tagetes daucoides, Tagetes elliptica Sm, Tagetes elongata, Tagetes epapposa, Tagetes erecta L, Tagetes filifolia Lag, Tagetes foeniculacea, Tagetes foetidissima, Tagetes hartwegii Greenm, Tagetes iltisiana Rydb, Tagetes inclusa, Tagetes lacera, Tagetes laxa Cabrera, Tagetes lemmonii, Tagetes linifolia, Tagetes lucida, Tagetes mandonii, Tagetes mendocina, Tagetes micrantha, Tagetes microglossa, Tagetes minima, Tagetes minuta L, Tagetes moorei, Tagetes mulleri, Tagetes multiflora Kunth, Tagetes nelsonii, Tagetes oaxacana, Tagetes osteni, Tagetes palmeri, Tagetes parryi, Tagetes perezii, Tagetes praetermissa, Tagetes pringlei, Tagetes pusilla, Tagetes riojana, Tagetes rupestris, Tagetes stenophylla, Tagetes subulata, Tagetes subvillosa, Tagetes tenuifolia, Tagetes terniflora Kunth, Tagetes triradiata, Tagetes verticillata, Tagetes zypaquirensis.

Cultivars

The African Marigold (*Tagetes erecta*)

The African marigolds are generally tall (up to 90 cm) with large sized double globular flowers of lemon, yellow, golden yellow, primrose, orange or bright yellow colours. There are also dwarf cultivars (20 to 30 cm) having large double flowers. The important cultivars are: Pusa Narangi Gainda, Pusa Basanti Gainda, Giant Double African Orange, Giant Double African Yellow, Cracker Jack, Climax, Dubloon, Golden Age, Chrysanthemum Charm, Crown of Gold, Spun Gold, Alaska, Apricot, Burpee's Miracle, Burpee's White, Cracker Jack, Crown of Gold, Double Eagle, Fluffy Ruffles, Fire Glow Gerldine, Glitters, Gaint Sunset, Golden Age, Golden Climax Gaint, Golden Jubilee, Golden Mammothmum, Golden Yellow, Goldsmith, Guinea Gold, Happiness, Hawaii, Honey Comb, Man of the Moon, Marry Helen, Mr Moonlight, Orange Fluffy, Orange Jubilee, Orangemum, Primrose, Sovereign, River Side, Sun Giants, Super Chief Double, Sutton's Gaint Orange Double, Texas, Yellow Climax, Yellow Fluffy and Yellowstone, etc.

French Marigold (*Tagetes patula*)

The French marigolds are mostly dwarf, early flowering and compact with dainty single or double blooms, borne freely and almost covering the entire plant. The colour flowers may be yellow, orange, golden yellow, primrose, mahogany, rusty red, tangerine or deep scarlet or a combination of these colours. The important cultivars are: Red Borcade, Rusty Red, Butter Scotch, Valencia, Sussana. However, in the market mostly orange colour cultivars are preferred and the cultivar which is dominating is African Giant Double Orange.

Single

Dainty Marietta, Legion of Honour, Naughty Marietta, sunny, Tetra Ruffled Red, etc.

Double

Bolero, Bonita, Brownie Scout, Burpee's Gold Nugget, Burpee's Red and Gold, Butterscotch, Carmen, Cupid Yellow, Eldorado, Fiesta, Goldie, Gypsy Dwarf Double, Harmony, Lemon Drop, Melody, Midget Harmony, Orange Flame, Petite Gold, Petite Harmony, Petite Orange, Petite Spray, Petite White, Petite Yellow, Primrose Climax, Red Brocade, Rusty Red, Spanish Brocade, Spray, Spun Yellow and Tangerine Yellow Pigmy.

F1 hybrids

Apollo, Climax, First Lady, Gold Lady, Gray Lady, Moon Shot, Orange Lady, Showboat (3x, sterile), Toreador, Inca Yellow, Inca Gold and Inca Orange.

Interspecific Hybrids

Burpee's Gold, Red Glow, Red Gold and Yellow Nugget.

Commercial Uses

Fresh Flowers

Freshly harvested flowers of African and French marigolds are used for religious offerings and social functions. The flowers remain fresh for 4-5 days at room temperature. Commercial growing for loose flower production is being done in Uttar Pradesh, Bihar, Delhi, Haryana, Punjab, Rajasthan, Madhya Pradesh, West Bengal, Maharashtra, Karnataka, Tamil Nadu and Andhra Pradesh. The cultivation of marigold yields a net profit of Rs. 1, 25,000 from the sale of fresh flowers from one hectare area, which is quite high as compared to many cereal crops.

Rich Carotenoid Source

Marigold is grown commercially for the extraction of carotene pigments xanthophylls are major carotenoid fraction in the flower petals. Marigold carotenoids are the major source of pigment for poultry feed. The pigment is added to the feed for the intensification of yellow colours of the egg yolk and broiler skin. Flower petals of African marigold are the best source of carotene for colouring foodstuffs. In India, the extraction of carotenoids is being done commercially in Kerala, Karnataka and Andhra Pradesh, primarily in Cochin, Hyderabad and Bangalore, respectively and it is being exported to Mexico regularly. Consequently large areas in Karnataka, Andhra Pradesh and Maharashtra are under contract cultivation of marigold for extraction of pigment.

Essential Oil Source

France, Kenya and Australia are the main marigold oil producing countries in the world.

Therapeutic Uses

Besides, flower production, marigold has very good medicinal value. Floral extract is used as a remedy for various eye diseases and ulcers. The juice extracted from petals and heated with an equal quality of ghee is given thrice daily as a remedy against bleeding piles. Flower extract purifies the blood.

Medicinal Uses

Marigold is used for stomach upset, ulcers, menstrual period problems, eye infections, inflammations and for wound healing. It is antiseptic. If the Marigold flower is rubbed on the affected part, it brings relief in pain and swelling caused by a wasp or bee. A lotion made from the flowers is most useful for sprains and wounds and a water distilled from them is good the sore eyes. The infusion of the freshly gathered flowers is beneficial in fever. Marigold flowers are mostly in demand of children ailment. Externally it is used in the treatment of alopecia. Internally it is used to treat bladder and kidney problems, blood in the urine, uterine bleeding and many more.

Other Uses

Bright yellow and orange marigold flowers are used to make garlands. They are even used to decorate the religious places. The leaves of its flowers are used as salads. Yellow dye has also been extracted from the flower by boiling. The burning herb repels insects and flies. Pigments in the marigold are sometimes extracted and used as the food colouring for humans and livestock. It is offered to the God and Goddess on the Durga Puja.

Preparation of Nursery Beds and Sowing of the Seeds

Raised nursery beds of convenient size (normally 1m wide and raised to 15 cm) should be prepared by added well rotten cow dung manure and sand may be mixed only when soil is heavy. To minimize the mortality of the plants in the nursery, the soil should be drenched with 0.2 per cent Captan before sowing. Seed sowing is adjusted in such a way that the prevailing temperature suits the seed germination. Seed germination occurs within 5-10 days of sowing at a temperature ranging from 18° to 30°C. Seeds rate per hectare is from 800 g to 1 kg. As soon as the beds are ready, seeds should be shown in lines spaced at 6 to 8 cm. Seeds may be sown at a depth of about 2 cm and then covered with a thin layer of leaf mould or sand or FYM. Watering is to be done daily with a watering cane. Thick sowing of seeds should be avoided because dense population results in week seedlings which take longer time to regain growth. For summer season crop, the seeds are sown in the second week of February while for rainy season crop, the sowing is done in the first week of June. Second week of September is the best time for sowing for the seeds for taking winter season crop.

Planting

The seedlings are transplanted in the field after 1 month of seed sowing. A spacing 40X40 cm for african marigold and 30x30 cm for french marigold have been found to be optimum for commercial cultivation. During summer and rains transplanting should be done in the afternoon or during cloudy weather. The terminal portion of the plants should be pinched as and when plants establish well in the field. This will help in the production of more side shoots.

Manures

Since marigold is fast growing crop, it requires high dose of nitrogen and moderate level of phosphorous and potash for better root development and quality flowers. Therefore, 40-50t/ha of FYM should be applied at the time of land preparation. In addition to FYM, it is advisable to apply 200 kg/ha nitrogen and 80 kg/ha each of phosphorous and potash for good flower yield. The full dose of phosphorous and potash should be given before the transplanting while nitrogen is given in two split doses (30 and 60 days after transplanting). It will be better if two foliar spray of 0.2 per cent urea are done at an interval of 15 days.

Irrigation

During the hot months of summer, the plants should be irrigated at an interval of 4-5 days to keep the soil moist and cool. Rainy season crop is irrigated as per the atmospheric conditions. Winter crop needs watering at 8-10 days intervals depending upon the water requirements of the plants.

Weeding and Hoeing

Weeds should be removed as soon as they appear. Weeds should never be allowed to set seeds otherwise they will continue menacing the crop. To create conducive soil conditions for plant growth, hoeing should be done at short intervals to keep the soil porous and to enable light, air and water to reach the roots easily to remove moisture retention capacity and to remove weeds. Hoeing should be done carefully so that the surface of stem or root is not damaged.

Flowering

In summer season crop, the flowering commences by the middle of May with maximum intensity in the month of June and continues till the onset of rains. After the onset of rains, the plants grow more vegetatively. Rainy season crop will start flowering by the middle of September and the flowering will continue till December. Flowering in winter crop will start by the middle of January and will continue till March.

Harvesting and Packing

The flowers are harvested when they are fully open. After harvesting, the flowers are packed in small and big bamboo baskets. They are then transported to nearest or distant markets by buses or Lorries. It is better to harvest the flowers in the morning and to store them in a cool place before packing. Different means of transportation *viz.*, Rickshaws, Buses, Trains are used to carry the flowers to markets depending upon the distance.

Flower Yield

If proper cultural practices are followed, it is not difficult to obtain high flower yield. A yield of 20-22 tonnes of fresh flowers can be obtained from one hectare of plantation in case of african marigold, while from french marigold it is 10-12 tonnes per hectare.

Pests and Diseases

Pests Management

Red Spider Mite (Tetranychu spp.)

These mites sometime appear on the plant near flowering time. Plants give dusty appearance. It can be controlled by spraying Metasystox 25 E C or Rogor or Nuvacron 40 E C or Kethone @ 1 ml/l of water.

Hairy Caterpillar (Diacrisia oblique)

This caterpillar eats away foliage. This caterpillar can be controlled by Nuvan or Thiodan at 1 ml/l of water.

Disease Management

Damping Off

It is caused by *Rhizoctonia solani* and appears as brown necrotic spots, girdling the radicle, later on extends to plumule and pre-emergence mortality. When infected seedlings are pulled, the root system appears fully or partially decayed. Seeds should be treated with Captan @ 3 g or Carbendazim @ 2.5 g per kilogram of seeds before sowing.

Collar Rot

It is caused by a number of pathogens and common ones are *Phytophtora sp., Rhizoctonia solani, Pythium sp.* Collar rot is caused either in nursery or in grown up plants. It can be prevented by soil sterilization or by using healthy seedlings.

Flower Bud Rot

It is caused by *Alternaria dianthi*. The disease mainly appears on young flower buds and results in dry rotting of buds. Symptoms are less prominent on mature buds but these buds fail to open. To control this disease regular spraying of the crop with Dithane M-45 @ 0.2 per cent should be followed.

Powdery Mildew

Oidium sp. causes powdery mildew in marigold. Whitish, tiny, superficial spots appear on leaves, later on the whole aerial parts of the plant is covered with whitish powder. The disease can be controlled by spraying with Karathane (40 E C) @ 0.5 per cent or dusting with sulphur powder at fortnightly intervals.

14

Orchids

Family Orchidaceae is a diverse and widespread family of flowering plants, with blooms that are often colourful and often fragrant, commonly known as the orchid family. Along with the *Asteraceae*, they are one of the two largest families of flowering plants, with between 21,950 and 26,049 currently accepted species, found in 880 genera. Regardless, the number of orchid species nearly equals the number of bony fishes and more than twice the number of bird species and about four times the number of mammal species. The family also encompasses about 6–11 per cent of all seed plants. The largest genera are *Bulbophyllum* (2,000 species), *Epidendrum* (1,500 species), *Dendrobium* (1,400 species) and *Pleurothallis* (1,000 species). The family also includes *Vanilla*, *Orchis* and many commonly cultivated plants such as *Phalaenopsis* and *Cattleya*. Moreover, since the introduction of tropical species into cultivation in the 19th century, horticulturists have produced more than 100,000 hybrids and cultivars. Orchids are the most sought-after flowers. They are expensive, elegant, dainty and rare to find, possess a miraculous beauty touch about them. Every woman longs for a man who will bring her orchids. Of course not every could

afford to bring orchids on their way home to their lady every day but even a one-time give away of an orchid signifies feelings of love, romance and miraculous feelings a man holds towards a woman. Eurasia, America and Canada are known as the homes of orchid but today they are found almost in every part of the world except for the arid zones. However, there are few arid areas that support the growth of orchids but very few species are native to such locations. This is where it belongs to, coming to the origination of orchid, its first species, comes a long history, attached and associated with different ways of treating, keeping, growing and discovering an orchid flower. A French botanist, Noel Barnard, was the first to discover this magnificent flower. Back in the year 1899, he came across seeds of a terrestrial orchid, Neottia. He discovered that these seeds had germinated in some fruits which were infected by mould. He collected it, thereby realizing the significance of it and produced a hybrid of orchid called Laelio cattleya with the help of pod parent. He further experiment it with fungus, sucrose and mixture of salep and produced more hybrids of orchids, marking a breakthrough foundation for the American history, taken forward by Dr. Lewis Knudson, an American scientist.

Another discovery was by Charles Darwin. He noted that a single orchid plant, specifically *Orphys maculata* (Origin: Europe), produced enough seeds in one season. The amount is too much to exactly calculate but he discovered that this will continue till the day the whole earth is filled with orchids. This is how one orchid gives birth to several and it becomes a continuous process. Pollination is an important process for the orchid. It was discovered later that it is done so by the insects as they carry it from one orchid to the other and then help in the fertilizing as well. There were even instances in Darwin's book on the orchids' pollination that this one specific orchid, Star of Bethlehem must be pollinated by the moths found in Madagascar. Knowing that orchid have a tremendous scent but we are unfortunate to smell it, insects are wildly attracted to it. Most white orchids are pollinated by night-flying moths, colourful orchids usually by bees and the smaller orchids by mosquitoes and coming down to the fact that such an important process as pollination is done via the help of specific insects; this could be one of the reasons why it is rare and expensive. Once the insect is lured to the flower, colorful markings and ridges on the lip serve as guides to lead the insect towards the nectary, causing the insect's head to come into contact with the stigma.The insect, when lured by the scent of the orchid, they sit and fall into the liquid glands of the orchid. In an effort to escape, they pick up pollinia on the way and head out from the lip of the orchid. As the orchid start to bloom in the season and tries to come out, the pollinia are deposited on it.

Orchid said to be the most beautiful of the God's creation,has conquered the cut flower industry all over the world during the past few decades.Valued for their exquisite flower and long keeping quality orchid keep fascinating people and are expected to do so for years to come. In India orchids are known since Vedic period. There are reference to orchid in ancient Sanskrit literature like Nighantus and Amarakosha.The term 'Vanda',applied to one of the widely cultivated orchid genera of Asian countries,derives from the Sanskrit term 'Vandika'.Orchid from about nine per cent of our flora.Nearly 1300 species of orchids spread over 140 genera are reported from our country. The term 'Orchid' had its origin form the Greek word 'Orchid' meaning testicles, referring to the paired underground tuber of terrestrial orchids. Before their aesthetic qualities were exploited,they were grown for their tubers, which were considered as aphrodisiac.The compilation known as Sishatha Samhita and Charaka Samhita (600-200 BC) also contain rheumatism in older days, in our country.Hanbenaria a terrestrial genus of orchid is also considered having medicinal value.

Botanical Description

Orchids are easily distinguished from other plants, as they share some very evident shared derived characteristics, or "apomorphies". Among these are: bilateral symmetry of the flower (zygomorphism), many resupinate flowers, a nearly always highly modified petal (*labellum*), fused stamens and carpels, and extremely small seeds.

Importance and Uses

Orchids with their bewildering range of flowers and beautiful colour combination provide a source of profound aesthetic pleasure to both owners and visitors. They are the most beautiful items for indoor decoration. A flowering plant, kept in a living room, remain fresh for many days and attract the attention of visitor. Ladies also use orchids flowers for the adornment of their hair. Orchids flower for the plant sale as well as cut flower production has developed into vast industries in many countries and sale of flowers run into million of dollars. The stem of the dendrobium species are used in making baskets in the philipines, Indonesia and New guinea. A few species of orchids have been considered to induce the female sterility among the women, valued for cut flower production and as potted plants.

Species and Cultivar

The family orchidiae consist 600-800 genera and 30,000-35000 spp. distributed in the throughtout the world. In India 1300 species are scattered all over north east Himalayas (600 spp.), north west (Himalayas 300 spp.), Maharashtra (130), Andaman and Nicobar Island (70 species) and western ghat 200 species. Genus *Cymbidum* is now among the top 10 cut flower of the world market where as *Dendrobium* the most widely cultivated tropical orchids is also marching a head to find a better place in the export market.

The genous *Cymbidum* is mainly grown in the north eastern Himalyan region, Sikkim, Assam and Tropical orchids is mainly confined to Kerala and some part of Tamilnadu. North eastern Himalayan region and the west –coast of Kerala are the main centre of production of orchids. The commercial orchids are both terrestrial and epiphytic with an abundance in epiphytic. All orchids are perennial herbs that lack any permanent woody structure. They can grow according to two patterns:

Monopodial

The stem grows from a single bud, leaves are added from the apex each year and the stem grows longer accordingly. The stem of orchids with a monopodial growth can reach several metres in length, as in *Vanda* and *Vanilla*.

Sympodial

Sympodial orchids have a front (the newest growth) and a back (the oldest growth). The plant produces a series of adjacent shoots which grow to a certain size, bloom and then stop growing and are replaced. Sympodial orchids grow laterally rather than vertically, following the surface of their support. The growth continues by development of new leads, with their own leaves and roots, sprouting from or next to those of the previous year as in *Cattleya,* while a new lead is developing, the rhizome may start its growth again from a so-called 'eye', an undeveloped bud, thereby, branching. Sympodial orchids may have visible pseudobulbs joined Terrestrial orchids may be rhizomatous or form corms or tubers. The root caps of terrestrial orchids are smooth and white. Some sympodial terrestrial orchids, such as *Orchis* and *Ophrys,* have two subterranean tuberous roots. One is used as a food reserve for winter periods and provides for the development of the other one, from which visible growth develops.

In warm and constantly humid climates, many terrestrial orchids do not need pseudobulbs. Epiphytic orchids, those that grow upon a support, have modified aerial roots that can sometimes be a few meters long. In the older parts of the roots, a modified spongy epidermis, called velamen, has the function to absorb humidity. It is made of dead cells and can have a silvery-grey, white or brown appearance. In some orchids, the velamen includes spongy and fibrous bodies near the passage cells, called tilosomes. The cells of the root epidermis grow at a right angle to the axis of the root to allow them to get a firm grasp on their support. Nutrients mainly come from animal droppings and other organic detritus collecting among on their supporting surfaces. The base of the stem of sympodial epiphytes, or in some species essentially the entire stem, may be thickened to form a pseudobulb that contains nutrients and water for drier periods. The pseudobulb has a smooth surface with lengthwise grooves and can have different shapes, often conical or oblong. Its size is very variable; in some small species of *Bulbophyllum,* it is no longer than two millimeters, while in the largest orchid in the world, *Grammatophyllum speciosum* (giant orchid), it can reach three meters. Some *Dendrobium* species have long, canelike pseudobulbs with short, rounded leaves over the whole length; some other orchids have hidden or extremely small pseudobulbs, completely included inside the leaves. With ageing, the pseudobulb sheds its leaves and becomes dormant. At this stage it is often called a backbulb. Backbulbs still hold nutrition for the plant but then a pseudobulb usually takes over exploiting the last reserves accumulated in the backbulb which eventually dies off. A pseudobulb typically lives for about five years. Orchids without noticeable pseudobulbs are also said to have growths, an individual component of a sympodial plant.

Leaves

Like most monocots, orchids generally have simple leaves with parallel veins, although some Vanilloideae have a reticulate venation. Leaves may be ovate, lanceolate or orbiculate and very variable in size on the individual plant. Their characteristics are often diagnostic. They are normally alternate on the stem, often folded lengthwise along the centre ("plicate"), and have no stipules. Orchid leaves often have siliceous bodies called stegmata in the vascular bundle sheaths (not present in the Orchidoideae) and are fibrous. The structure of the leaves corresponds to the specific habitat of the plant. Species that typically bask in sunlight, or grow on sites which can be occasionally very dry, have thick, leathery leaves and the laminae are covered by a waxy cuticle to retain their necessary water supply. Shade-loving species, on the other hand, have long, thin leaves. The leaves of most orchids are perennial, that is, they live for several years, while others, especially those with plicate leaves as in *Catasetum*, shed them annually and develop new leaves together with new pseudobulbs. The leaves of some orchids are considered ornamental. The leaves of the *Macodes sanderiana*, a semiterrestrial or rock-hugging ("lithophyte") orchid, show a sparkling silver and gold veining on a light green background. The cordate leaves of *Psychopsis limminghei* are light brownish-green with maroon-puce markings, created by flower pigments. The attractive mottle of the leaves of lady's slippers from tropical and subtropical Asia (*Paphiopedilum*), is caused by uneven distribution of chlorophyll. Also, *Phalaenopsis schilleriana* is a pastel pink orchid with leaves spotted dark green and light green. The jewel orchid (*Ludisia discolor*) is grown more for its colourful leaves than its white flowers. Some orchids, as *Dendrophylax lindenii* (ghost orchid), *Aphyllorchis* and *Taeniophyllum* depend on their green roots for photosynthesis and lack normally developed leaves, as do all of the heterotrophic species. Orchids of the genus *Corallorhiza* (coralroot orchids) lack leaves altogether and instead wrap their roots around the roots of mature trees and use specialized fungi to harvest sugars.

Flowers

Orchidaceae are well known for the many structural variations in their flowers. Some orchids have single flowers but most have a racemose inflorescence, sometimes with a large number of flowers. The flowering stem can be basal, that is, produced from the base of the tuber, like in *Cymbidium*, apical, meaning it grows from the apex of the main stem, like in *Cattleya*, or axillary, from the leaf axil, as in *Vanda*. As an apomorphy of the clade, orchid flowers are primitively zygomorphic (bilaterally symmetrical), although in some genera like *Mormodes*, *Ludisia* and *Macodes*, this kind of symmetry may be difficult to notice. The orchid flower, like most flowers of monocots, has two whorls of sterile elements. The outer whorl has three sepals and the inner whorl has three petals. The sepals are usually very similar to the petals (*tepals*) but may be completely distinct. The medial petal, called the *labellum* or lip which is always modified and enlarged, is actually the *upper* medial petal; however, as the flower develops, the inferior ovary (7) or the pedicel usually rotates 180 degrees, so that the labellum arrives at the lower part of the flower, thus becoming suitable to form a platform for pollinators. This characteristic, called resupination, occurs primitively in the family and is considered apomorphic, a derived characteristic all Orchidaceae share. The torsion of the ovary is very evident from the longitudinal section shown (*below right*). Some orchids have secondarily lost this resupination, *e.g. Zygopetalum* and *Epidendrum secundum*. The normal form of the sepals can be found in *Cattleya*, where they form a triangle. In *Paphiopedilum* (Venus slippers), the lower two sepals are fused into a synsepal, while the lip has taken the form of a slipper. In *Masdevallia*, all the sepals are fused. Orchid flowers with abnormal numbers of petals or lips are called peloric. Peloria is a genetic trait but its expression is environmentally influenced and may appear random. Orchid flowers primitively had three stamens but this situation is now limited to the genus *Neuwiedia*. *Apostasia* and the Cypripedioideae have two stamens, the central one being sterile and reduced to a staminode. All of the other orchids, the clade called *Monandria*, retain only the central stamen, the others being reduced to staminodes. The filaments of the stamens are always adnate (fused) to the style to form cylindrical structure called the gynostemium or column. In the primitive Apostasioideae, this fusion is only partial; in the Vanilloideae, it is more deep; in Orchidoideae and Epidendroideae, it is total. The stigma is very asymmetrical, as all of its lobes are bent towards the centre of the flower and lie on the bottom of the column. Pollen is released as single grains, like in most other plants, in the Apostasioideae, Cypripedioideae and Vanilloideae. In the other subfamilies, that comprise the great majority of orchids, the anther, carries and two pollinia. A pollinium is a waxy mass of pollen grains held together by the glue-like alkaloid viscin, containing both cellulosic strands and mucopolysaccharides. Each pollinium is connected to a filament which can take the form of a caudicle, as in *Dactylorhiza* or *Habenaria*, or a *stipe*, as in *Vanda*. Caudicles or stipes hold the pollinia to the viscidium, a sticky pad which sticks the pollinia to the body of pollinators. At the upper edge of the stigma of single-anthered orchids, in front of the anther cap, there is the rostellum, a slender extension involved in the complex pollination mechanism. As mentioned, the ovary is always inferior (located behind the flower). It is three-carpelate and one or, more rarely, three-partitioned, with parietal placentation (axile in the Apostasioideae).

Fruits and Seeds

The ovary typically develops into a capsule that is dehiscent by three or six longitudinal slits, while remaining closed at both ends. The seeds are generally almost microscopic and very numerous, in some species over a million per capsule. After ripening, they blow off like dust particles or spores. They lack endosperm and must enter symbiotic relationships with various mycorrhizal basidiomyceteous fungi that provide them the necessary nutrients to germinate, so that all orchid species are mycoheterotrophic during germination and reliant upon fungi to complete their lifecycles. As the chance for a seed to meet a suitable fungus is very small, only a minute fraction of all the seeds released grow into adult plants. In cultivation, germination typically takes weeks. Horticultural techniques have been devised for germinating orchid seeds on an artificial nutrient medium, eliminating the requirement of the fungus for germination and greatly aiding the propagation of ornamental orchids. The usual medium for the sowing of orchids in artificial conditions is agar agar gel combined with a carbohydrate energy source. The carbohydrate source can be combinations of discrete sugars or can be derived from other sources such as banana, pineapple, peach or even tomato puree or coconut water. After the preparation of the agar agar medium it is poured into test tubes or jars which are then autoclaved to sterilize the medium. After cooking, the medium begins to gel as it cools.

Pollination

The complex mechanisms which orchids have evolved to achieve cross-pollination were investigated by Charles Darwin and described in *Fertilisation of*

Orchids. Orchids have developed highly specialized pollination systems, thus the chances of being pollinated are often scarce, so orchid flowers usually remain receptive for very long periods, rendering unpollinated flowers long-lasting in cultivation. Most orchids deliver pollen in a single mass. Each time pollination succeeds, thousands of ovules can be fertilized. Pollinators are often visually attracted by the shape and colours of the labellum. However, some *Bulbophyllum* species attract male fruit flies (*Bactrocera* spp.) solely via a floral chemical which simultaneously acts as a floral reward (*e.g.* methyl eugenol, raspberry ketone or zingerone) to perform pollination. The flowers may produce attractive odours. Although absent in most species, nectar may be produced in a spur of the labellum or on the point of the sepals or in the septa of the ovary, the most typical position amongst the Asparagales. In orchids that produce pollinia, pollination happens as some variant of the following sequence: when the pollinator enters into the flower, it touches a viscidium, which promptly sticks to its body, generally on the head or abdomen. While leaving the flower, it pulls the pollinium out of the anther, as it is connected to the viscidium by the caudicle or stipe. The caudicle then bends and the pollinium is moved forwards and downwards. When the pollinator enters another flower of the same species, the pollinium has taken such position that it will stick to the stigma of the second flower, just below the rostellum, pollinating it. The possessors of orchids may be able to reproduce the process with a pencil, small paintbrush, or other similar device. Some orchids mainly or totally rely on self-pollination, especially in colder regions where pollinators are particularly rare. The caudicles may dry up if the flower has not been visited by any pollinator and the pollinia then fall directly on the stigma. Otherwise, the anther may rotate and then enter the stigma cavity of the flower (as in *Holcoglossum amesianum*). The slipper orchid (*Paphiopedilum parishii*) reproduces by self-fertilization. This occurs when the anther changes from a solid to a liquid state and directly contacts the stigma surface without the aid of any pollinating agent or floral assembly. The labellum of the Cypripedioideae is poke-bonnet-shaped and has the function of trapping visiting insects. The only exit leads to the anthers that deposit pollen on the visitor. In some extremely specialized orchids such as the Eurasian genus *Ophrys*, the labellum is adapted to have a colour, shape and odour which attracts male insects via mimicry of a receptive female. Pollination happens as the insect attempts to mate with flowers. Many neotropical orchids are pollinated by male orchid bees which visit the flowers to gather volatile chemicals they require to synthesize pheromonal attractants. Each type of orchid places the pollinia on a different body part of a different species of bee so as to enforce proper cross-pollination. A rare achlorophyllous saprophytic orchid growing entirely underground in Australia, *Rhizanthella slateri*, is never exposed to light and depends on ants and other terrestrial insects to pollinate it. *Catasetum*, a genus discussed briefly by Darwin, actually launches its viscid pollinia with explosive force when an insect touches a seta, knocking the pollinator off the flower. After pollination, the sepals and petals fade and wilt but they usually remain attached to the ovary.

Asexual Reproduction

Some species such as *Phalaenopsis*, *Dendrobium* and *Vanda*, produce offshoots or plantlets formed from one of the nodes along the stem through the accumulation of growth hormones at that point. These shoots are known as *keiki*.

The Primary Classification

The family orchideae is divided into several genera and species. They are further divided based on their habit (terrestrial, epiphytic) and growth habit (monopodial, sympodial). Most of the commercially important orchids are epiphytic in nature.

Commercially Important Genera

The genera of orchids which could be commercially grown under the condition prevailing in India may be classified into two categories namely the monopodial orchids (arachnis, vanda, phalaenopsis) and the sympodial orchids (dendrobium, cymbidium, cattelya, oncidium).

Cultivars

New pink, Hieng beauty, Emma white, Kasem white, Sonia-28, Kioni beauty, Banyot pink, Hawailan beauty+kasem pink, Sonia -17, White nern, Boonchoo gold, Kanjana green, Jay swee king+jacquline, Jacquline Thomas, Madam vipor, Nette white, Pink tips, Banyat pink, Sakura pink, Candy stripe, Pramott-11, Sabine.

Propagation

Conventional Methods

The orchid can be propagated by both seed or by vegetative means. Plant raised from seeds take a long time to bloom. The vegetative means can be broadly grouped into two, the conventional method such as by division, cutting, airlayering and the tissue culture method even the under the traditional method of propagation, the procedure is not the same for monopodial and sympodial.

Propagation of Monpodial Orchids

Stem Cutting

About 40-50 cm long top cutting with at least two well developed aerial roots are ideal. Intermediate cutting can be used. If smaller cutting, 2-3 nodes are used the time taken to flowering will be longer.

Flower Stalk Cutting

These are also reported to be usefull in certain genera like ex- phalaenopsis, phaius, calanthe and thunia. Some time vegetative shoots are produced from the flowered spikes in the genera. otherwise the flower stalks can be cut off and laid horizontally on moist media like sphagun moss or coconut husk bits.

Layering

Airlayering was found to give success in mompodial like vanda. For this a slanting cut may be given on the stem, at about 20-30 cm below the apex. The pound may covered with some moist media as in the case of air layering, the layers can be separated and planet when they strike roots.

Propagation of Sympodial Orchids

Division

The method involves division of large clumps into smaller unites. This is the most common method of propagation in sympodial orchids. This method is especially suitable for catteya, dendrobium, cymbidium, epidendrum.

Off-shoots or Keikis

Orchids like dendrobium sometimes produces small plants with roots at the node of psedudobulbs.these are called keikkis meaning babies. These when sufficiently grown are not to be separated carefully from the mother plant and potted independently.

Backbulbs

The older shoot or canes of sympodial orchids which are lesser active physiologically are called backbulbs. These backbulbs may be severed off the mother plant and kept horizontally over a moist medium after they will strike roots and sprout from nodal region. Then they can be separated and planted in individual pots.

Environmental Factor

Since majority of the commercial orchids are epiphytic and that they are grown in media other then soil. The physical and chemical properties of soil do not influence directly the growth and development of orchids.

Light

Of all the weather element essential for proper growth and development of orchid, Light play the most important role besides having its direct effect on plant function such as photosynthesis. It influence the temperature and relative humidity.

Temperature

Temperature is largely a function of sunlight. It regulate the growth and flowering orchids considerably. The cultivated orchids classification as a warm tropical intermediate (subtropical) and cool (temperate) species is based on their preference to a particularly range of temperature. One important point, however, is that orchid should never be subjected to sudden changes in temperature.

Humidity

This refers to the amount of water vaper in the atmosphere. Orchids, in general prefer high humidity for their growth and flowering. The ideal range varies with genera and species. Monopodial types reguire higher humidity than sympodial ones.

Air Circulation

Orchids thrive well when the atmosphere is fresh. Air circulation not only keeps the atmosphere clean and fresh but also helps to keep both temperature and humidity at desired levels. These ultimately help to bring down the incidence of pests and diseases. Orchids can be divided horticulturally in two broad groups.

Terrestrial

Grown on the ground *e.g.*, cymbidium, paphioledilum.

Epiphytes

Grow on the trees *e.g.*, vanda, dendrobium.

Media for Orchids

Media Character

It should have good air action. Drainage should not absorb much water and should not degenerate easily broken bricks, gravel, tile bites charcoal bits and tree fern are component of media used for growing epiphytic orchids. The component of media for growing epiphytic orchids. The component are washed thoroughly before filling in pots for terristial orchids a judicious mixture of humus leaf mould, dried mannure chopped fern and sphagnum mass sufficient for example for epiphytic orchids. The pot is filled with media and the plants are placed near the edge of the pot at the growing point.

Media for Terrestrial Orchids

Porosus and rich humus added with decayed leaves from the media for terrestrial orchids found in forest floors. Simulating this natural medium, apooling medium containing equal parts of leaf mould, soil and sand would be desirable. Clay soil, bonemeal, saw dust, charcoal dust, manure, wood shaving are also used in various proportion to satisfy the need of terrestrial orchids.

Media for Epiphytic Orchids

In nature, the epiphytic orchid spread their roots over the branches of tree exposing them fully. The type of media used for growing epiphytic orchids should provide a surface over which the plant can cling to since the main quality sought for in the media for epiphytic orchids is to supply moisture. The main character of the media should be retention of adequate moisture for a sufficiently long period of time.

Common Component of Media

Rockwool

Rockwoll is an inert material that looks like dirty cotton. There are 2 varieties of rockwool: one absorbs water, the other repels water. Ten years ago or so we used water absorbent rockwool as part of our potting mix but gave up on it because it holds too much water and was difficult to mix. But the worse was its propensity to collapse which reduced the air space in the potting material.

Tree Fern Fibre

Tree fern from the roots of a fern called "tree fern". It is an excellent material that is relatively easy to use and will easily stay fresh for 3 years. The material is relatively expensive. There are 2 varieties of tree fern: one is sort of light brown and flexible, is mostly available in Hawaii and is called Hawaiian tree fern, also known as "hapu", the other is dark brown and rigid and comes mostly from Central America. The latter is what is typically referred to as "tree fern" and, unless otherwise noted, when we mention "tree fern" we mean the rigid one from Central America. Medium tree fern holds just about as much water as medium fir bark (fine tree fern holds more than fine fir bark), but by far not as much as sphagnum moss. It is more expensive than orchid bark. Tree fern also comes in 3 sizes: fine (or seedling grade), medium and coarse

Osmunda Fibre

Osmunda fiber was a choice material in decades past. Nowadays it is not as readily available and it is expensive. Furthermore, it comes in relatively large chunks that you have to cut into about 1.2 cm chunks in order to use it. Great exercise for the wrist but who has the time. Furthermore, when using osmunda fiber you must make sure the fibers are aligned vertically so as to allow the water to drain.

Bark

Orchid bark is an excellent material. It is easy to use, it will not hold excessive water and under normal use will not need to be refreshed for about 2 years. Orchid bark (which usually is the bark from redwood or Douglas fir), comes in 3 sizes (sizes are also referred to as "grades"): small size (also known as "seedling" size), medium and large (or coarse) size. The sizes used are mostly the seedling size and the medium size. Bark is rarely used alone. Most growers add to it one or more of the following: perlite, sponge rock (which is expanded perlite), charcoal (horticultural grade), sphagnum moss, tree fern, peat moss, Perlite and sponge rock are used to create more air space in the mix. Charcoal is used to absorb harmful materials that may be in the water. Sphagnum or peat moss is used to increase the water holding capacity of the mix.

Coconut Husk

Coconut fiber is not widely used to grow orchids. It is used a thin layer of it to line Vanda and wire baskets to prevent other potting materials from falling out of the basket. It is used to "stuff " Vanda baskets when potting vandaceous orchids (Vandas and related) so as to hold just a little bit of moisture and to help to hold the plant in place. Use it sparingly and "fluff " it because if it is too dense it will stay wet and Vandaceous orchids hate this

Peat Moss

It is not used as the main ingredient nor as an additive. It has a high water retention capacity and it does stay relatively intact for several years. Some growers include it as part of their mix.

Charcoal

In the US charcoal is rarely used as the main or sole ingredient of the potting material. Some growers add charcoal to their potting material because charcoal absorbs toxins that may be present in the water and released by the roots of plants.

Common Container

Eartern or Clay Pots

Plastic Potes

Plastic pots Most commercial growers use plastic pots because they are inexpensive, they are lighter and they are easier to store because they hold water for a longer period than other pots/containers, because mineral salts (from water and fertilizer) will not adhere to them and because roots will not get attached to them. Plastic pots are excellent containers for growing orchids. Their only draw back, is that some plants, notably Dendrobiums, might get top heavy in them. Green plastic pots are the ones most commonly used plastic pots. Lately clear plastic pots have become more widely available. Advocates of clear plastic pots claim the light transmission of clear plastic pots enable roots to photosynthesize. If use plastic pots, look for pots with a fair number of drainage holes (4 to 8 holes on 7.6 cm to 10.2 cm pots, 8 to 12 holes on 12.2 cm to 15.2 cm pots).

Baskets

Vanda baskets are used mostly for Vandas and vandaceous orchids, but can be used for most orchid genera. Most Vanda baskets are made of cedar or teak wood. Unfortunately the cedar baskets available today tend to decay in a couple of years. Teak baskets are expensive and because teak trees are being depleted it is not ecologically friendly to buy teak baskets. Fortunately plastic Vanda baskets have made their appearance. The ones (10.2 cm and 20.3 cm) are made of sturdy plastic that should last forever if it recycle them.

Slabs

Cork slabs are used for mounting orchids. Pieces of cork range can be as small as 5.1 cm by 7.6 cm or as large as 30.5 cm by 61.0 cm. Some orchids will only thrive when mounted on a piece of cork or on a tree fern slab or a piece of driftwood but many that grow fine in pots will also thrive on a piece of cork and it makes for a much more natural and interesting.

Commercial Planting

Spacing-Cutting at 20 - 30 cm

Manures and Fertilizers

A fertilizer complex containing nitrogen, phosphorus and potash in equal proportion like 17-17-17 complex is ideal for general application. The concentration may be adjusted between 0.2 and 1.0 per cent depending upon the situation already explained. Organic manurs like cow dung, Neem oil cake and poultry manure are also used for orchids. These are found to be more ideal to monopodial types. These may be soaked in water for 4-5 days for fermentation. It is diluted to 10-15 times with water filtered and sprayed over the plant.

Repotting

Each orchid genus has different requirements for potting media. It is very important to have the correct medium for each type of orchid, depending on whether it is terrestrial or epiphytic-tree dwelling. Growing media commonly include fir bark, coconut husk, sphagnum moss, tree fern fibers and perlite and frequently a mixture of two or three of these materials. All orchids potted in a typical bark medium need to be repotted every 18 to 24 months, depending on the needs of the individual plant. The primary purpose of repotting is to provide fresh media, not necessarily a larger pot but pot size should be selected according to the size of the root mass. Orchids like to be a little tight in their pots. Orchids transferred to overly large pots tend to concentrate their energy on root growth and may not show new growth or foliage for several months. Orchids may be potted in plastic, clay or decorator pots and the type of pot selected may influence watering frequency; plants in clay pots will need more frequent watering as they will dry out a little faster. Always select pots with drainage holes; orchid roots in contact with standing water will rot and die, killing the plant. Media in the center of larger pots may remain wet for long periods and become an unhealthy environment for roots. This can be avoided by placing pieces of broken terracotta in the bottom of the pot. A smaller pot inverted into a larger one can also help with drainage and aeration with the roots of the plant draped over and around the smaller pot. Some orchids, such as Phalaenopsis have roots capable of photosynthesis. For these plants clear pots have become popular as they allow light to get to the roots.

Irrigation

Beside improving the humidity of the environment, application of water should also in washing out the deposits of fertilizer residues which may otherwise be harmful to the plants. In summer, two irrigation are ideal. A heavy irrigation in the morning, followed by a dry period and a light irrigation in the late afternoon is preferred.

Harvesting

The spike of the harvested when a few bud on the top remain unopen under tropical condition. Harvesting when a few buds on the top remain unopen under tropical conditions, harvesting during early morning or evening is preferred. Harvested spikes also known as stem are immediately put in a bucket of water. The cut end is fully immersed. These are then taken to the packing house for grading and packing the cut ends of the spikes wrapped with wet cotton and tied with a rubber band alternatively. The cut ends are inserted into a plastic tube containing water. Number of buds to be retained on the spike and grading varies in different places. It is usually 25-50 per cent of the total flowers in a spike. In a spike producing larger number of flowers as in dendrobiums, those having less then 5 flowers are not usually preferred. Dendrobium is harvested while two or three buds are still unopened, it add to the charm of the spike. This applies to all other orchids which produce long spike like phalaenopsis, aranda and archinis. The number of unopened buds to be retained may be lesser in smaller spike.

Yield

Yield of the spike varies from genus to genus and cultivar to cultivar. On an average 6-8 spikes are available from commercial cultivar of dendrobium. In sonia 17, more number of spike are produced in Kerala for a period of 2-3 months with the onset of south –west monsoon, there is a decline in flowering in most of the cultivars.

Post-harvest Management

Shipping and Packaging

Everything from rough handling to extreme temperatures can damage orchids in transit. Some nurseries rush through packing and shipping, which can lead to a disaster for the plants. The properly stake the plant and protect both leaves and flower spikes should the plant send you in spike. The use of shredded wax paper and/or tissue or cotton-like padding to prevent injury. Carefully wrap the plant in newspaper and if the weather is either too cold or hot one will insulate your plants by lining the box with styrofoam planks and or micro foam wrapped around the news paper. Fill the voids of the box with styrofoam peanuts for additional cushioning and insulating value. In cold weather one will include non toxic chemical heat packs to keep your orchids warm for up to two days. The stalk are usually packed in cartones. The size of the carton is determined by the genera and the number of spike to be packed. Generally 50-100 spike are packed in a carton. The cartons are to be provided with sufficient number of holes all around to facilitate good aeration.

Orchid Grading

One supply our customers with only the best quality orchids, giving them our Grade-A classification. It means healthy orchids in prime condition. One has lovingly cared for our orchids for many months in our tropical glasshouses prior to their despatch to their new homes. One has been fed with a mix of nutrients made specifically for our orchids and watered according to seasonal light levels and ambient air temperatures. Orchids have glossy, rich green leaves that are generally blemish-free. Even 'polish' them before they leave our glasshouses, using an environmentally friendly leaf shine. Orchids have green, robust stems with healthy nodes. Stems will grow in a variety of directions as is their customary habit in their natural environment. However, one stake the orchids in an upright position using wooden orchid sticks and small, plastic clips. Orchids will have a number of aerial roots growing out of their pots. In their native habitat of tropical rainforests, the roots act as anchors, attaching themselves to tree branches. Plenty of aerial roots is a sign of the orchid's health and vigour. Orchids are either double or multi stemmed plants with numerous buds and flowers. In order for you to enjoy your Grade A orchids for many months to come, one send our orchids to their new homes with as many buds and as few open flowers as possibly can. The grades of orchids are described underneath:

Grade A	40 – 50cm in height	Double stemmedglossy leaves, generally blemish free.	12 - 16 closed buds and open flowers.
Grade B	40cm in height	Single stem, glossy leaves, minor blemishes.	6 - 8 buds and open flowers.
Grade C	30cm in height	Single stem, glossy leaves, some blemishes.	4 - 6 buds and open flowers.

Major Insects/Pests

Table 14.1: Symptoms and Management of Various Insects of Orchids

Insects/Pests	*Symptoms*	*Management*
Scale	The young and mature scale suck the juice from the leaves, petiole and shoots by injecting toxins into the plant.	Spray of fish oil rosin soap 2 per cent or methyl parathion 0.05 per cent or dimethoate 0.05 per cent
Mealy bug	They such the sap from the plant parts and secrete honey dew.	Spray methyl parathion 0.04 per cent
Thrips	Both adult and nymph suck the cell sap from leaves and growing bud affected leaves curl while petals develop streaks with scorched margin dry.	Dimethoate or Monocrotophos at 0.05 per cent at 10 days interval

Major Diseases

Table 14.2: Symptoms and Management of Various Diseases of Orchids

Diseases	*Symptoms*	*Management*
Leaf spot	Sunkun spot appear at any place on the leaves which letter turn brown.	Bordeaux mixture 6:6:50 or 4:4:50,Dithane M-45 0.2 per cent and Bavistin 0.1 per cent
Heart rot	The leaves of the affected plant turn yellow and drop off pesudobulbs have dark rotted area inside.	Use Dithane M-45 and Metalaxy at 0.2 per cent.
Root rot	Disease kill the seed ling and retard the growth of mature plants	Use Dithane M-45 0.2 per cent
Flower blight	The infection causes small necrotic spots and ultimate rotting of flowers.	Use Carbendazim 0.2 per cent

15

Rose

Rose is a popular garden flower throughout the world owing to its exquisite shape, size, colour and delightful fragrance. It has been referred in Sanskrit literature as Tarunipushpa, Atimanjula and Simantika. It was adorned both in ashrams of saints and in royal palaces. Its cultivation developed with distillation of roses. The interest has from then on increased and at present it is grown for garden display, exhibition and commercial purposes. The total area under commercial production is around 4000 ha in our country and the major producing regions are Maharastra, Karnataka, Punjab, U.P., Delhi and Chandigarh, while in M.P., Gujarat, Tamil Nadu, West Bengal, Rajasthan, they are cultivated to a limited extent.

Cultivars

The rose breeding work in India is of comparatively a recent origin and was taken up in right earnest in the last five decades. A rose nurseryman from Deoghar (Bihar) is acclaimed the pioneer rose breeder in India who evolved a fairly large number of roses. Several outstanding cultivars like Dr. Homi Bhabha, Delhi Princess and Banjaran created the real intrest in rose breeding. The evolution of over 100 rose cultivars were made. More than 300 cultivars have been developed in India. Some important cultivars developed at the IARI are Anurag, Arjun, Bhim, Chitchor, Poornima, Raja Surendra Singh of Nalagadh, Dr. B.P. Pal, Dr. Homi Bhabha, Jawahar, Kanakangi, Mechak, Mridula, Mrinalini, Raktagandha among hybrid tea group and Banjaran, Chandrama, Delhi Princess, Loree, Mohini, Nellabari, Prema, Rupali, Sadabahar, Shabnam, Sindoo, Suchitra, Suryakiran among the Floribunda group. Some important cultivars bred by amatures and nurserymen are Srinivasa, Sugandha, Raja Ram Mohan Roy and Dr. Radha Krishnan in hybrid tea group. About 30 cultivars have also been developed in polyantha, mininature and climber groups. Several cultivars have been evolved through natural mutation or as bud sport of the existing cultivars. At IARI, 3 rose cultivars were developed through induced mutation. These are Abhisarika from Kiss of Fire, Pusa Christina from Christian Dior and Madhosh from Ceulzar.

Soil and Climatic Requirements.

For successful rose cultivation, soil should be loamy having sufficient organic matter with a pH of 6.0 to 7.5. Proper drainage is to be ensured as water stagnation affects rose growing. In heavy soils, sand and gravel may be mixed.Bright sunshine for the whole day or atleast in the forenoon is must for good growth and flowering of roses. The plants should be free from shades of trees and protected from strong winds. Generally, the prevailing climatic conditions of our country is found to be suitable for its cultivation. They are widely cultivated in northern and eastern plains of India, Pune and Nasik areas of Western India and also in mild climate of Bangalore.

Environmental Factors

Intensity and duration of light are found to influence growth and flower production. Plants growing under shade produced thinner leaves with low chlorophyll content than plants grown in full sunlight. A reduction of 12 per cent light intensity resulted in 14 per cent less flower production in Sonia and Illona. In polyantha roses cv. Charles, Bonnet grown under 16 hrs light period produced largest number of flowers. Most of the rose cultivars are grown at a night temperature of 15.5°C to obtain good quality and quantity of roses. In Sonia, a night temperature of 15°C and day temperature of 22°C are found to be optimum to obtain higher yields. Higher humidity is found to increase the incidence of pests and diseases.

Propagation

Roses are mainly propagated through budding for commercial cultivation in India. Attempts have also been reported to propagate them through cuttings. In advanced countries, tissue culture technique is used. In the budding – technique, T-budding is commonly practiced. The time of budding varies from place to place and the right stage of budding is when plants have good sap flow and the cambium tissue is highly active. The best time for budding in eastern India is from January-March. In north India, it is done from December to February. In Bangalore, budding is done almost throughout the year but best time will be in March-April. Two rootstocks are commonly used. In North India, *Rosa indica* is used while in South India, it is *Rosa multiflora*. At IIHR, a 'thornless' rootstock has been obtained as a natural mutant from *Rosa multiflora inermis*. Its performance has been *at par* with *Rosa multiflora*. Recently, in Maharastra, propagation through polybag method is gaining popularity. At U.A.S. Bangalore, propagation through girdling and etiolation, through treating hard wood and semihard wood cuttings with IBA 2000 ppm, has been successful. At I.I.H.R. cv. Queen Elizabeth is propagated successfully by treating the cuttings with IBA, 2000 ppm by quick-dip method.

Time and Method of Planting

Initially the land should be dug to a depth of 30 cm and should be thoroughly pulverized. This is followed by removal of stones, gravel, brick pieces and grass roots. This is done during summer when soil is well exposed to sun, air and also monsoon, so that soil settles down before planting. In general, roses can be planted any time of the year except during very hot summer and during heavy rains. The ideal months are October to December in most parts of the country. Pits of 30 cm × 30 cm are prepared. The plants are lowered into the pit such that bud union is 5.0 cm above the soil level. On lowering, the pit is filled with the same dug out top soil and pressed to anchor the plant firmly.

Spacing

The planting density/unit area plays an important role in production of quality flowers. A spacing of 60-75 cm × 60-75 cm (13096-16308 plant/hectare) has been recommended. Studies on high density planting of 60 × 30 cm (29652 plants/hectare) and 30 × 30 cm (44478 plants/hectare) have shown that higher yields can be obtained per unit area.

Nutrition

Basic manuring with farm yard manure at the rate of 6-8 kg/bush (74.73 tonnes/ hectare) is advocated. Supplementary manuring with oil cakes or cowdung slurry, poultry or sheep droppings can also taken up. Generally a fertilizer dose of 20-40g N, 20g P_2O_5 and 20g K_2O/sq.m. or 223:48:148 kg/ha is to be applied after pruning. The dose may be repeated after the first flush of flowering is over. 494.2 kg CAN, 370.65 kg Super phosphate, 247.1 kg of muriute of potash/ha at the time of pruning is to be applied. A second dose of 75 kg CAN is to be applied after 30 days of pruning and 3rd dose in the first week of February. In alluvial soil zone of west Bengal a fertilizer dose of 600 kg N, 200 kg P_2O_5 and 200 kg K_2O/ha has been recommended. At I.I.H.R. for high density planting a combination of 247.1 kg N, 494.2 Kg P_2O_5 and 370.65 kg K_2O/ ha have resulted in highest flower production. Foliar feeding of rose plants helps in quick absorption of nutrients and is an additional method of improving rose plants especially under adverse soil conditions. The use of 2.5 g mixture of urea and potassium dihydrogen phosphate mixed in equal proportions by weight and made upto one litre of solution is recommended. Whenever, the soils are deficient of micro nutrients, it is recommended to get the symptoms properly identified before spraying micronutrients. A mixture containing 15 g manganese sulphate, 20g magnesium sulphate, 10 g chelated iron and 5 g borax have been found to be very effective for obtaining brighter colour in flowers and foliage. The spraying should not be done when the plants are in full bloom.

Growth Regulators

Gibberellic acid has proved effective in regulating growth and flowering of roses. Spraying GA_3 four times at monthly intervals at 250 ppm increased the number of flowers, thickness and freshsheet of flower stems. Two sprayings with B-9 at 4000 ppm have resulted in more branching, early flowering and increased flower size.

Pruning

Pruning is an important operation for maintaining floriferousness and flower quality along with the vigour of the rose plants. It ensures the removal of unproductive growth and keeps the bush in proper shape and size. In most parts of our country, pruning is done once a year during October-November and it has been suggested to stagger prunings at weekly intervals from 23rd September to 16th October to get a regular supply of flowers throughout winter from December to March in a commercial garden. In Bangalore generally pruning is done twice a year *i.e.*, in June and November, although some commercial growers prune once a year. In each bush while pruning, 4-5 canes are retained with 4 buds on each cane. After pruning, either Blitox paste is applied or fungicide or insecticide spray is given. There are three types of pruning, Light, moderate and hard. In light pruning diseased and unwanted portions of the stem are removed and healthy shoots are cut either at the second or third bud immediately below the flower bearing footstalk. In moderate pruning, main branches are cut at about half the length of growth. In Floribundas, light and moderate pruning is advised for production of abundant flowers thus enhancing mass effect in the bed. Hard pruning involves cutting back the main shoots to 4 buds from the base as in Hybrid Teas.

Harvesting and Yield

Harvesting is done either before sunrise or in the late afternoon to avoid high temperature during the day. Flowers harvested for the local market or for loose flower purpose are generally cut when one or two of the outer petals begin to unfurl. Flowers selected for distant market or for export should be harvested at tight bud stage to retain colour, freshness, withstand transportation stress and also to last longer in vases. While harvesting flowers cut should be made such that two five-leaflet leaves remain below the cut. Immediately after cutting, the flowers have to be placed in a bucket of fresh and clean water upto the flower base. The delay will result in vascular blockage due to air entry. After this the flower stalks should be recut 2 cms above the previous cut end. If they are not immediately used, they can be stored at air temperature of 4.4-7.2°C for 6-12 hrs. It is necessary to dispatch flowers within 24 hrs of harvest and reach the destination within 24 hrs to 30 hrs, preferably in early morning when most flower transactions take place. About 3.71 lakh flowers/ha/annum can be obtained from a well maintained gardens.

Greenhouse Production and Growing Roses for Export

Attempts are made in India recently to produce flowers under protected cultivation. However, it is still on a low key. Presently area under protected cultivation is around 5-6 ha. In Netherlands, the area under cover is 10 times as much and about 8000 blooms are obtained per day from an area of 12,000 sq.m. The cost of a glasshouse of this size is approximately 3.7 million. The following criteria are to be borne in mind for exporting of roses. The roses for export should exclusively be grown in glass house, fibre glass house or plastic house so that the production of flowers. meets the quality requirements. For internal consumption as well as for short distance foreign markets like, Middle east countries, roses may be grown out doors or under partial cover. The red, pink and orange colours are mostly preferred. There is less demand for yellow and white roses. The biocoloured roses are not preferred in the market. The length of the stem for large flowered cultivars is 60 to 90 cm while for small flowered cultivars it is 40 to 50 cm. The size of the large flowered bud is 3-3.5 cm and that of the small flowered is 2-2.5 cm. Large flowered cultivars suited for export are Sonia Meilland (Salmon pink), Illona (Red), Red success (Red), Baccara (Vermillion Red), Audlsmeer Gold (Golden yellow). Small flowered cultivars are Motrea (red), Mercedes (red), Red Garnettee (red), Belinda (copper-gold orange), Carol (deep rose-pink), Golden Times (deep yellow). For continuity of blooms, it will be necessary to prune bushes 3-4 times at weekly intervals. About 1,20,000 cut flowers of exportable quality can be obtained from one ha by following close planting (30×30 cm) and providing plastic cover over the plants during November-February. Rose flowers are cut while still in the bud stage after the sepals have curled back and the colour is fully showing. In small flowered cultivars and Floribundas, the flowers are cut just when they begin to open in the cluster. Rose flowers are packed in corrugated cardboard boxes of 120 cm long × 45 cm wide × 25 cm high. They should reach destination within 24-30 hours after harvest preferably early morning when most of the flower transaction takes place. Hence, rose plantations should be as close to the International Airport as possible.

Pests and Diseases

Roses are attacked by number of pests like red scales, aphids, thrips, red spider mites and chaffer beetles. A regular and rotational field spray with monocrotophos or rogor or metasystox at 1.5-2 ml/litre will control the insect damage. For red scales, granular insecticides like Carbofuran at 1.5 kg a.i./ha can be applied. Among diseases, Black spot, powdery mildew and die back are most commonly seen. Occurrence of powdery mildew is severe during summer while that of Black spot is severe during rainy season. The latter is controlled by spray of Dithane M-45 (2g/lit) and the former is controlled through spray of sulfex (2g/lit) at ten days interval. Both diseases are best controlled by preventive measures like burning of infected portions and spraying of fungicides.

16

Tuberose

Tuberose (*Polianthes tuberosa* L.)is one of the most important tropical ornamental bulbous flowering plants cultivated for production of long lasting flower spikes. It is popularly known as Rajanigandha or Nishigandha. It belongs to the family Amaryllidaceae and is native of Mexico. Tuberose is an important commercial cut as well as loose flower crop due to pleasant fragrance, longer vase-life of spikes, higher returns and wide adaptability to varied climate and soil. They are valued much by the aesthetic world for their beauty and fragrance. The flowers are attractive and elegant in appearance with sweet fragrance. It has long been cherished for the aromatic oils extracted from its fragrant white flowers. Tuberose blooms throughout the year and its clustered spikes are rich in fragrance; florets are star shaped, waxy and loosely arranged on spike that can reach up to 30 to 45 cm in length.The flower is very popular for its strong fragrance and its essential oil is important component of high- grade perfumes. 'Single' cultivars are more fragrant than 'Double' type and contain 0.08 to 0.14 per cent concrete which is used in high grade perfumes. There is high demand for tuberose concrete and absolute in international markets which fetch a very good price. Flowers of the Single type (singlerow of perianth) are commonly used for extraction of essential oil, loose flowers, making garland, while that of Double cultivars (more than two rows of perianth) are used as cut flowers, garden display and interior decoration. Fragrance of flowers is very sweet, floral and honey like and can help give emotional strength. The flower spike of tuberose remains fresh for long time and finds a distinct place in the flower markets. Due to its immense export potential, cultivation of tuberose is gaining momentum day by day our country. The Tuberose (*Polianthes tuberosa*) is a perennial plant related to the agaves, extracts of which are used as a note in perfumery. The common name derives from the Latin *tuberosa,* meaning swollen or tuberous in reference to its root system. *Polianthes* means "many flowers" in Greek. In Mexican Spanish, the flower is called *nardo* or *vara de San Jose* which means "St. Joseph's staff". This plant is called as Rajnigandha in India which means 'fragrant at night'. The tuberose is a night-blooming plant native to Mexico, as is every other known species of *Polianthes*. It grows in elongated spikes up to 45 cm long that produce clusters of fragrant waxy white flowers that bloom from the bottom towards the top of the spike. It has long, bright green leaves clustered at the base of the plant and smaller, clasping leaves along the stem. Epiphyllous adhesion of stamens is seen in the flower. Members of the closely related genus *Manfreda* are often called "tuberoses". In the Philippines, the plant is also known as *azucena* and while once associated with funerals, it is nowTuberose is endemic to Central America and southern Mexico. The flower landed in India via the Philippines. From this migration two other routes were traced to Europe.(One was through Spain to the Languedoc in southwestern France and the other, through Persia, traveling through Liguria to Provence.

Long before the botanical name *Tuberosa polianthes*, it was named *Hyacinthus indicus tuberosa radice* in 1601, literally "hyacinth of India." Today, literally, it means "flower of towns." The term "polianthes" comes from Greek *polis* meaning "city" and *anthos*, "flower." The name of its genus *Tuberosa* characterizes the fleshy tuberous root. Tuberosa is from the agave family, the genus comprises a total of 13 species and each is more or less fragrant. Nowadays, tuberose is mainly cultivated in Morocco, The Comoros, India and China for processing. Very fragrant tuberose has been in many legends and stories that reveal its bewitching nature. The best-known story during the Italian Renaissance, where it was forbidden for unmarried girls to walk through the gardens where tuberose exercised her erotic and intoxicating power, so they would not succumb to drunkenness and men maddened by the erotic smell. If jasmine absolute reveals the joy on the faces of the people who smell it, it was said that a woman who exudes the scent of tuberose cause mimicry recalling orgasm. In 1530, a French missionary imported from Mexico the first tuberose bulbs and secretly grew them in the garden of a monastery near Toulon. The Tuberose was developed then in Languedoc, in Italian Liguria and Provence through the second half of the seventeenth century, for glove makers merchants and apothecaries. The tuberose flower is toxic and inedible like the lily, lily of the valley or narcissus. Two varieties of *Polianthes tuberosa* are used. "Usual tuberose" is used in perfumery as it provides better organoleptic characteristics.(Another flower variety, neatly called "pearl tuberose" is used for floral decorations.(In Latin America, pearl tuberose is a prominent part of bridal bouquets and is deposited at the entrance of the bride's house to distinguish it. It is grown in indoor gardens and patios to welcome with its fragrance. In India, women make wreaths and necklaces of this flower called *Maalai* for temples as offerings. Tuberose concrete is part of the composition in many incense mixtures and cosmetics.

Uses of Tuberose

Essential Oil

The health benefits of Tuberose Essential Oil can be attributed to its properties as an aphrodisiac, deodorant, relaxing, sedative and warming substance.Tuberose is not a very popular name in the world of herbal medicines. In fact, it is very popular and priced among perfume manufacturers. Its flower has a beautiful fragrance, which is active at night, which is the only time that this flower blooms. Tuberose is popularly known as "Night Queen", "Mistress of the Night" or "Raat ki Raani" as it known in Hindi. It grows well in Central America and India and is in high demand in the countries of the Indian Subcontinent, Middle East, and Africa in order to make perfume.Its scientific name is *Polianthes tuberosa* and its essential oil is extracted from its flowers by a solvent extraction method. The main components of Tuberose Essential Oil are benzyl alcohol, butyric acid, eugenol, farnesol, geraniol, menthyl benzoate, menthyl anthranilate and nerol.

Medicinal Uses of Essential Oil

Aphrodisiac

Justifying its romantic and sensual names that include "Night Queen" and "Mistress of the Night", this oil behaves as an aphrodisiac. It is commonly used as an aphrodisiac both in aromatherapy as well as in herbal medicines. The very strong, intense, and intoxicating floral fragrance made with the Essential Oil of Tuberose fills the atmosphere with romance and builds an atmosphere of love. This oil is found to be very effective in treating frigidity and a lack of libido. Certain components of this essential oil stimulate those parts of the brain that are responsible for arousal, sexual feelings and libido both when used in aromatherapy or taken orally. It also relaxes the mind, which is a pre-requisite for successful sex, as tension and stress are two of its biggest enemies. Furthermore, it has a warming effect on the organs because it increases circulation of the blood which helps cure erectile dysfunctions and impotency.

Deodorant

This essential oil, which is world famous for its use in perfumes certainly doesn't need any explanation about its function as a deodorant. The rich, intense and long-lasting floral fragrance is an ideal choice for a deodorant, which is why it is so popular in countries with hot and humid climates, as they have to frequently deal with sweat and the resultant body odour.

Relaxing

The pleasant fragrance and various chemical components of this oil have relaxing effects on the brain, nerves and the muscles. It calms people and gives relief from stress, tension, anxiety, depression, anger, nervous afflictions, convulsions, cramps, spasmodic coughs and diarrhea.

Sedative

This essential oil is good for sedating inflammations, particularly those pertaining to the nervous system and the respiratory system. However, to have this sedating effect, it should be used in relatively high dilution.

Warming

Tuberose essential oil stimulates and increases blood circulation throughout the body thereby inducing a warming effect. This effect counters the feeling of cold in winters, keeps the respiratory system warm, prevents the deposition of phlegm and catarrh, increases activity and also helps to cure sexual disorders.

Other Benefits

This essential oil can also be used to keep the skin free of infections and cracks, in-hair oils to counter nausea and reduces the tendency of vomiting and the effects of foul smells in certain areas.

Tuberose Absolute Oil

Tuberose absolute is called as Rajnigandha in Ayurveda meaning 'night fragrance'. The flowers of this plant bloom during the night time and release their scent for attracting the attention of nocturnal moths for pollination. Just like Jasmine, tuberose also has a strong floral scent that is effective even hours after plucking it. Essential oils have been used for their aroma and therapeutic values by priests, devotees and worshipers to draw the attention of Gods, Goddesses and Spirits to ward off evil powers and to sanctify the places of worships. In this way, the aroma of tuberose absolute oil with Venus as the guiding planet is said to be the blessed oil to Maya who is respected as the origin of Gods. This oil was also used in the primordial times for opening the heart chakra, capturing compassion, vision, love, psychic sensitivity, creativity and trance work.

Others Uses

Tuberose can successfully be grown in pots, borders, beds and commercially cultivated for its various uses. The flowers of tuberose are used for making artistic garlands, floral ornaments, bouquets, buttonholes, gajras and extraction of essential oil. It is a popular cut flower, not only for use in arrangements but also for the individual florets that can provide fragrance to bouquets and boutonnieres. The long flower spikes are excellent as cut flowers for table decoration. The flowers emit a delightful fragrance. Tuberose represents sensuality and is used in aromatherapy for its ability to open the heart and calm the nerves, restoring joy, peace and harmony.Tuberose flowers have long been used in perfumery as a source of essential oils and aroma compounds. Tuberose oil is used in high value perfumes and cosmetic products.

Furthermore, fragrant flowers are added along with stimulants or sedatives to the favourite beverage prepared from chocolate and served either cold or hot as desired. Tuberose bulbs contain an alkaloid lycorine, which causes vomiting. The bulbs are rubbed with turmeric and butter and applied as a paste over red pimples of infants. Dried tuberose bulbs in powdered form are used as a remedy for gonorrhoea. In Java, the flowers are eaten along with the juices of the vegetables.

Area and Distribution

Tuberose is grown commercially in a number of countries including India, Kenya, Mexico, Morocco, France, Italy, Hawaii, South Africa, Taiwan, North Carolina, USA, Egypt, China and many other tropical and subtropical areas in the world. In India, commercial cultivation of tuberose is popular in Bagnan, Kolaghat, Midnapur, Panskura, Ranaghat, Krishnanagar of West Bengal; Coimbatore and Madurai districts of Tamilnadu; Pune, Nashik, Ahmednagar, Thane, Sangli of Maharashtra; East Godavari, Guntur, Chitoor, Krishna District of Andhra Pradesh; Mysore, Tumkur, Kolar, Belgaum and Devanhalli taluk in Karnataka; Guwahati and Jorhat in Assam; Udaipur, Ajmer and Jaipur in Rajasthan; Navsari and Valsad of Gujarat and parts of Uttar Pradesh and Punjab. As per area and production statistics of National Horticulture Board (2013), the total area under tuberose cultivation in the country is about 7. 95 lakh hectare. The production of loose and cut flowers is estimated to be 27.71 '000 MT and 1560.70 lakh, respectively.

Species and Cultivars

There are about fifteen species under the genus Polianthes, of which twelve species have been reported from Mexico and Central America.Of these, nine species have white flowers, one is white tinged with red and two are red. Except *Polianthes tuberosa* L., all the others are found growing wild.

Cultivars

There are four types of tuberoses named on the basis of the number of rows of petals they bear. They are Single, Semi-double, Double and Variegated.

Single

They bear pure white flowers with one row/whorl of corolla segment. Flowers are highly scented and are extensively used for loose flower purpose, essential oil and concrete extraction. Single types are more fragrant than double. Concrete content has been observed to be 0.08 to 0.11 per cent. Loose flowers are used for making floral ornaments. Its floral buds are greenish white. The per cent seed setting is high in single. Single Mexican, Kalyani Single, Shringar, Prajwal, Arka Nirantara, Rajat Rekha, Hyderabad Single, Calcutta Single, Phule Rajani, Kahikuchi Single, Pune Single are main cultivars.

Description of some Important Single Cultivars

- ☆ **Arka Nirantara:** Arka Nirantara is released by Indian Institute of Horticultural Research (IIHR), Bangalore. It has white, single flowers with prolonged blooming.

- ☆ **Shringar:** This tuberose hybrid has been developed from a cross between 'Single x Double' and was released by Indian Institute of Horticultural Research (IIHR), Bangalore. It bears single type fragrant flowers on strong and sturdy, medium spikes. The flower buds are attractive with slightly pinkish tinge. The spikes have more number of flowers and the individual florets are larger and appealing compared to the local 'single' cultivar. Loose flowers of this hybrid can be used for garlands and for extraction of tuberose concrete.The spikes can be used as cut flower. The loose flower yield of this hybrid is about 36 per cent higher than the existing local single cultivar. Yield of loose flowers is about 15,000 kg/ha per year which is 40 per cent higher than 'Calcutta or Mexican Single' and the concrete content of the hybrid is at par with Mexican Single. Shringar is preferred by farmers and perfumery industries. This hybrid is tolerant to root knot nematodes (*Meloidogyne incognita*).
- ☆ **Prajwal:** This hybrid which bears single type flowers on tall stiff spikes is a cross between 'Shringar' x 'Mexican Single'.The hybrid was released by Indian Institute of Horticultural Research (IIHR), Bangalore. The flower buds are slightly pinkish in colour, while the flowers are white. The individual florets are large in size, compared to 'Local Single'. It yields twenty per cent more loose flowers than 'Shringar'. It is recommended both for loose flower and cut flower purpose.
- ☆ **Single Mexican:** It is a single flowered cultivar. This cultivar produces maximum flowers is considered as lean months for tuberose flowers yield.

Semi-Double

They consist of flowers bearing 2-3 rows of corolla segments on straight spikes.

Double

They bear flowers having more than three rows of corolla segments. It is a single flowered cultivar. This on straight spikes. Flower colour is white and also tinged with pinkish red. The main cultivars are Pearl for Double, Kalyani Double, Swarna Rekha, Hyderabad Double, Culcutta Double, Vaibhav and Suvasini.

Description of some Important Double Cultivars

- ☆ **Suvasini:** It is a double flowered multi whorled cultivars released by Indian Institute of Horticultural Research (IIHR), Bangalore. It is a cross between 'Single' and 'Double'. This cultivar produces more number of flowers per spike. The spikes are best suited for cut flowers. This tuberose hybrid is multi-whorled with bold,large, pure white fragrant flowers borne on long spikes in contrast to off-white flowers of local cv. 'Double'. The number of flowers per spike is more and flower opening is uniform in this hybrid as compared to the local 'Double' cultivar. Spike yield is 26 per cent more compared to the local Pearl Double 'Double' cultivar. Spikes are best suited for cut flower purpose.

- ☆ **Vaibhav:** This hybrid which bears double flowers on medium spikes is from the cross 'Mexican Single' x IIHR – 2' and was released by Indian Institute of Horticultural Research (IIHR), Bangalore. The flower buds are greenish in colour in contrast to pinkish buds in 'Suvasini' and 'Local Double'. Flowers are white. Spike yield is 50 per cent higher compared to 'Suvasini'. Hence, recommended for cut flower purpose.
- ☆ **Pearl Double:** The flowers tinged with red in the 'Double' type are known as 'Pearl'. Pearl Double is high flower yielder with quality flowers. They are mainly used for cut flower and bouquet purpose as well as loose flower and for extraction of essential oil. Concrete recovery has been found to be 0.06 per cent. It does not open well and is not commercially viable as the single cultivar.
- ☆ **Variegated:** These are some streaked leaf forms, known as 'variegated'. In these cultivars, silvery white or golden yellow streaks are visible on leaves. National Botanical Research Institute, Lucknow has developed two variegated cultivars Rajat Rekha and Swarna Rekha by gamma irradiation.
- ☆ **Rajat Rekha:** It is a double flowered cultivar released by National Botanical Research Institute, Lucknow.The flowers are double with golden yellow steaks along the margins of leaf. It is a gamma ray induced mutant, in which mutation occurred in chlorophyll synthesis resulting in change in leaf colour. Concrete content has been found to be 0.062 per cent.
- ☆ **Swarna Rekha:** It is a double flowered cultivar released by National Botanical Research Institute, Lucknow.The flowers are double with golden yellow steaks along the margins of leaf. It is a gamma ray induced mutant in which mutation occurred in chlorophyll synthesis resulting in change in leaf colour. Concrete content has been found to be 0.062 per cent.

Climate and Soil

Climate

Tuberose prefers to grow in an open sunny location, away from the shade of trees. It requires warm and humid climate although flowering is profuse under mild climate. Under extremes of high (>40°C) or low temperatures the spike length and the quality of the flowers is severely affected. A temperature from 20-30°C is considered to ideal for this crop.

Soil

Tuberose can be grown on wide variety of soils ranging from light,sandy loam to a clay loam. It can also be successfully grown as a commercial crop even in those soils which are affected by salinity and alkalinity conditions if better agronomical practices are adopted. The soil should be at least 45 cm deep, well drained, friable, rich in organic matter and nutrients with plenty of moisture in it. Tuberose should grown in well drained place. Crop is sensitive to water stagnation it cannot tolerate water logging even for a short period. So essential to ensure proper drainage pests or else planting should be done on done on bund. Fertile, loamy and sandy soils having a pH in the range of 6.5 to 7.5 with good aeration and drainage are ideal for tuberose cultivation. A place protected from strong winds in the soil is preferable.

Propagation

Commercially tuberose is propagated by bulbs and division of bulbs. Selection of suitable bulbs is very important for successful cultivation. Spindle shaped bulbs free from diseases with average diameter of about 1.5-2.0 cm size or above should be preferred for planting. Tuberose bulbs have a definite rest period after lifting them from soil. Dipping the bulbs in 4 per cent solution of Thiourea can break the resting period. Propagation through mature bulbs is expensive; therefore, multiplication of growing stock can be done by division of bulbs. Large sized bulbs having 2.1 cm or more in diameter are suitable for planting purpose. The bulbs are cut into 2 to 3 vertical sections, each containing a bud and part of the basal plate. These sections are treated with fungicide and planted vertically in a rooting medium with their tips just showing above the surface. A moderately warm temperature should be maintained. New bulblets along with roots develop from the basal plate. At this stage, bulblets are transferred to the ground.

Land Preparation

The soil should be ploughed to a depth of 30-40 cm during January and exposed to sun for at least 15 days for killing weeds and insects. Well rotten FYM @20 tonnes/ha is incorporated with the soil immediately after ploughing. The soil is brought to fine tilth by breaking the clods and removing the weeds. The field is laid out into pots of convenient sizes with irrigation channels and ridges and furrows. Under sub-tropical conditions, the bulbs are planted in February-April. Well-developed spindle-shaped bulbs, with diameter 1.56 cm and above forming at the outer periphery of the clump, are considered ideal for planting. Freshly harvested tuberose bulbs can be used for planting 4-5 weeks after harvesting. Planting fresh bulbs leads to profuse vegetative growth and poor flowering. Recommended spacing for planting the bulbs is as follows:

Table 16.1: Spacing and Number of Bulbs per Hectare of Various States

State	*Spacing (cm)*	*No. of Bulbs/ha*
Maharashtra	15x20	3,30,000
West Bengal	25x25	1,60,000
Karnataka	30x22.5	1,48,000
Uttar Pradesh	30x30	1,11,000

The depth of planting varies between 3.0-7.0 cm depending upon the diameter of bulb and the soil type. Planting is deeper in sandy soil as compared to clay soil. In sandy loam, soil planting of bulbs is done at the depth of 6.0 cm. In general planting is done in such a way that the growing portion of the bulb is kept at the ground level.

Treatment of Bulb

Dipping the bulbs in 4 per cent solution of thiourea can break the resting period. Pre-plant storage of bulbs at 10°C for a period of 30 days will improve the plant growth, increase spike and flower yield. Pre-planting treatment of bulbs with GA_3, etherel or thiourea promotes early appearance of flower spike and produces longer spikes with maximum number of florets.The bulbs are first thoroughly cleaned and treated with Bavistin (0.2 per cent) for 30 minutes. Dipping the bulbs for about 20-30 minutes in a solution of Emisan (0.2 per cent), Thiram (0.3 per cent), Captan (0.2 per cent) or Benlate (0.2 per cent) is recommended. Dry in shade before planting or storing. Before planting treatment, bulbs in systemic fungicide and before storing in contact fungicide.

Manures and Fertilizers

Tuberose is a gross feeder and responds well to the application of organic and inorganic manures. Apart from FYM (20 tonnes/ha), a fertilizer dose of 200 kg of N, 200 kg P_2O_5 and 150 Kg K_2O per hectare is recommended, of which 100 kg N and entire quantity of P and K is applied as a basal dose. The balance N is given in two split doses at thirty days interval. The 8kg ZnSo4, 2kg Boron and 1 kg Sodium Molybdate enhance the quality and quantity of flower.

Use of Growth Regulators

The application of CCC at 5000ppm and GA at 1000ppm induce early flowering, increased flower stalk production and improves the quality of flowers.

Irrigation

Initially, irrigation is given immediately after planting in order to set them in the ground and to provide them with sufficient moisture for growth initiation. Subsequent irrigation is given depending upon the prevailing weather conditions. Usually during summer (April-June), it should be irrigated at weekly intervals and during winters at 10 days intervals.

Weed Control

The field should be kept clean by periodical weeding at monthly intervals. Manual weeding is generally practiced. Hoeing between plants at regular intervals is helpful in loosening the soil and uprooting weeds. Control of weeds by using chemicals is also found effective. Application of Alachlor @2.0kg/ha; Pendimenthalin @1.25kg/ha or Metachlor @2.0kg/ha significantly reduced the weed population.

Harvesting

In India tuberoses is cultivated for production of flower spikes and loose flowers on a commercial scale for the domestic market. Flowers are ready for harvest in about 3 months of planting. August - September is the peak period of flowering. For marketing of flower spikes, the tuberose is harvested by cutting the spikes from the base when 1-2 pairs of flowers open on the spike. Individual flowers which grow at the horizontal position on flowers stalk are picked in the early morning. The spikes are clipped by using a sharp knife/secateur that gives a clean cut, leaving about 4-6 basal portion of the scape so as not to damage the growing bulb.

Flower Yield

Flower yield varies with cultivars, plant density, bulb size at time of planting and crop management. After a two-year experiment with tuberose cv. single recorded flower yield of 15,000 kg per hectare or 2,75,000 spikes during first year. The average yield of first ratoon crop was obtained 18,500 kg or 3,25,000 spikes per hectare.

Post-Harvest Technology

Vase Life

Just after removal, the lower end of the spike should be immersed in water for prolonging the life of spikes. The spikes are made ready by removing the unwanted leaves to minimise the transpiration. Further, pulsing of spikes at low temperature (10°C), for about four hours with the ends immersed in water, is helpful in prolonging life of spikes to be sent to distant markets.

Grading

The flower spikes are graded according to the stalk length, length of rachis, number of flowers per spike and weight of spike. Straight and strong stem of uniform length and uniform stage of development are preferred. Flowers should be free from bruises and pests and diseases. Florets are graded according to their sizes for loose flowers.

Packaging

Loose flowers of single-flowered tuberose are packed in bamboo baskets covered with wet gunny bags. About 10-15 kg fresh flowers are packed in each basket and transported to the nearby wholesale market where they are sold by weight. The spikes are packed by wrapping them first in wet newsprint sheets and subsequently in corrugated sheets, making bundles of convenient sizes. Such bundles are finally packed in strong card boxes, which are quite handy.

Harvesting, Curing and Storage of Bulbs

Harvesting stage of tuberose bulb is important for storage of bulbs and their growth. The bulbs are harvested when the flowering is over and plant ceases to grow. At this stage, the old leaves become dry and bulbs are almost dormant. Irrigation is withheld and soil is allowed to dry before digging out the bulbs. After digging, the bulbs are lifted out, the bulblets are separated and used as seed stock for the next season. The bulbs are graded based on their size and are placed on shelves to dry or cure. The bulbs must be stored or have their position changed every few days to prevent fungal attack and rotting. Curing can also be done by tying the bulbs in bunches and hanging them on frames and walls.

Yield

Flowers production varies with cultivars and depends upon bulb size at planting time and density planting and cultural practices adopted. Tuberose planted at a spacing of 30×30 cm with a plant population of 1,11,000 plants/ha yield about 90,000 marketable spikes and 1.8 lakhs flowering size bulbs.

Ratoon Crop

After harvesting the main crop, the flower stalks are headed back (cut to the base) and the plots should be well manured and irrigated. About 3-4 ratoon crops can be taken from a single planting. For the proper growth and development of plants, fertilizer dose as given in the main crop should be applied in two equal split doses in January-February and April. All other cultural practices should be done as in case of main crop. There is early flowering in ratoon crop as compared to main crop. The ratoon crop results in more number of spikes but reduces number of florets, length of spikes and weight of flowers. Therefore, ratoon crop should be used only for loose flowers or oil extraction purpose. In temperate climate, during November-December, when temperature drops, leaves of the plants turn yellow and die and plants undergo dormancy. Digging of bulbs should be done at this stage.With the increase in temperature, the crop regains growth from the previously planted bulbs which is termed as ratooning. For ratooning in tuberose, the yellowing plants should be twisted from the ground level which leads to early maturing of bulbs.

Pests and Diseases Management

Pests

Bud Borer (*Helicoverpa armigera*)

This pest mainly damages flowers. Eggs are deposited singly on growing spikes. Eggs are deposited singly on growing spikes. Larvae bore into buds and flowers and feed on them by making holes. Collection and destruction of damaged buds reduces the damage. Setting up of light traps help to control population by attracting them. Sprays of Endosulphan, 0.07 per cent or Methyl Parathion, 0.05 per cent taken up at appearance of eggs on buds and tender foliage controls borer damage. Neem oil, 1 per cent also gives considerable protection by repelling various stages of pest.

Aphids

These are tiny insects, soft bodied, green, deep purple or black in colour. These usually occur in clusters and feed on flower buds and young leaves. Spraying the infected plants with Malathion @0.1 per cent at an interval of 15 days is effective.

Grasshoppers

These feed on young leaves and flower buds. Affected plants with damaged foliage and flowers lose their elegance, especially during rainy season. Dusting the plants with 5 per cent Cythione/Folidol dust may prevent the damage. Scraping of buds expose egg masses to natural enemies. Netting prevents damage from hoppers to nurseries. Spraying of Quinalphos @0.05 per cent or Malathion 0.1 per cent or Carbaryl @0.2 per cent protects foliage of newly transplanted crop.

Red Spider Mites

Mites thrive well under hot and dry conditions, usually on the undersides of the leaves, where these make webs, if allowed to continue. These are usually red or brown in colour and multiply fast. Mites suck sap which results in the formation of yellow strips and streaks on the foliage. In due course of time, leaves become yellow, silvery or bronze and distorted. Spraying with Kelthane @ 1.2 per cent concentration is effective to control the mites.

Rodents

Rodents do considerable damage to tuberose plants in the field by making burrows. Poison bait is quit helpful in checking rodent menace in the field. Commercial bait by the name, 'Roban' is available in the market and the same may be used effectively.

Thrips

Thrips feed on leaves, flower stalk and flowers. These suck the sap and damage the whole plant. Sometimes, these are associated with a contagious disease known as 'bunchy top' where the inflorescence is malformed. Thrips can be controlled by spraying the plant with 0.1 per cent Malathion

Weevils (*Myllocerus* sp.)

The weevils are nocturnal in habit and damaged shoots and leaves. Usually, they feed the edge of the leaves, producing a characteristic notched effect. Larvae feed on roots and tunnel into the bulbs. Applying BHC dust (10 per cent) in the soil before planting controls larvae.

Diseases

Stem Rot

The disease symptoms are preceded by the appearance of prominent spots of loose green colour due to rotting which extend and cover the entire leaf. The infected leaves get detached from the plant. More or less round sclerotic, brown spots are formed on and around the infected leaf. As a result, the infected plant becomes weak and unproductive. The disease can be controlled by soil application of Brassicol (20 per cent) @ 30kg/hectare.

Botrytis Spot and Blight (*Botrytis elliptic*)

The disease appears during the rainy season. Infected flowers show dark brown spots and ultimately the entire inflorescence dries up. The infection also occurs on the leaves and stalks. Spraying the plants with Carbendazim @2g/litre of water effectively controls the disease. The treatment should be repeated at 15 days interval.

Sclerotial Wilt (Sclerotium rolfsii)

The initial symptom of this disease is flaccidity and drooping of leaves. The leaves become yellow and dry. The fungus mainly affects the roots and the infection gradually spreads upward through the tuber and collar portion of the stem. Both tubers and roots show rotting symptoms. Thick cottony growth of the fungus is visible on the rotten stem and on petioles at the soil level. Drenching the soil with 0.3 per cent Zineb is effective in controlling the disease.

17

Zinnia

Zinnias are popular in flower gardens because of their variable coloured blooms and their ability to withstand hot summer temperatures. They are easy to grow from seeds. Plants come in many colours, shape and size adapted well to most growing conditions. Out of about twenty species, there are some eight to ten species of zinnia known to the garden but only few of them have become favourite and that abundantly deserves the pre-eminence. They have attained position as one of the most splendid of annual flowers. Most varieties are prolific bloomer, making them excellent for landscape. They are also available as compact, varieties suitable for hanging baskets, window boxes and containers. Recently, new hybrid zinnia cultivars have attracted attention as a cut flower crop. There are at least a hundred varieties in the diversity of flower colour and types. The colour of zinnias are white, cream, green, yellow, apricot, orange, red, bronze, crimson, purple and lilac with stripped, speckled and bicoloured flowers. An important variety of zinnia is Coccinea or Scarlet-rayed. This variety became popular after its existence in 1829 and was thought much of for its brilliant colour

and stately habit. During the fifty years that have elapsed since it appeared, the flower has been in all characters and now possess a race of perfectly double zinnias, the flower of which show no central disc but are perfect rosettes of equisite form and of every shade of colour except for blue. There is not a more striking instance of floral advancement, accomplished by systemic selection, than is afforded by zinnia, which is at once one of the largest and long standing of many good and cheap annual flowers.

Origin and Distribution

The zinnia is named in the honour of Dr. Johann Gottfried Zinn, Professor of botany and natural history at the University of Gottingen. Zinnia is originally from scrub and dry grasslands in an area stretching from the South-West America to South America but primarily Mexico.Zinnias are grown mostly as annual in Mexico, Texas, New Mexico, Colorado, Brazil and Chile. It is sometimes called the Mexican marigold as its designation in some degree is justified in case of the yellow varieties. In early days zinnias were not popular. Although native to North and South America, not everyone found them attractive or desirable as a landscape plant. In fact, when Spanish first saw the zinnia species in Mexico, they considered the flower so unattractive that they named it "mal de ojos" which in Spanish for "Sickness of Eye". Zinnia seeds brought back to Europe in eighteenth century caused little excitement although the plant did not get good name. But it was not until the late nineteenth century that plant breeders started to pay any attention to this species. The start of real popularity of the zinnia came around 1920 when Bodger Seeds Ltd. Introduced the dahlia flowers Giant Dahlia. John Bodger discovered it as a natural mutation in the field of Mammoth and within the next few years selected the California Giant (large, flat-flowered variety) from the strain. The most attractive feature was availability of flowers in different colours. It also was considered to be a new trend in plant habit and flower form. It even won a gold medal from the Royal Horticultural Society of England, making everyone want to grow this prize winning flower. The many garden form of *Zinnia elegans* developed from the wild plant that grows in Mexico. Common zinnia has escaped and naturalized in part of southern United States including Central Florida. Recently cultivation of new hybrid zinnias have started in some American and European countries under open as well as protected conditions for cut flower production.

Botanical Description

Zinnia is a genus of 20 species of annual and perennial plant of family Compositae. The wild form is a coarse, upright, bushy plant upto 20-100 cm high, with solitary daisy-like flower heads on long stems and opposite, sandpapery, lance shaped leaves. Zinnia leaves are opposite and usually stalkless, with a shape ranging from linear to ovate and pale to middle green in colour. The ray flowers are purple, the disc yellow and black and the entire head is about 5 cm across. Flowers are single, semi-double or double forms. Flowers are single or bicolours and with button type flowers, beehives (small blooms with rows of flat petals), cactus shaped flowers (petals roll under and twist and bend) and dahlia shaped flowers (large, flat and usually semi-double). Newer varieties are resistant to powdery mildew and other diseases. A large number of varieties have been developed from selection and hybridization worldwide.

Species

Most accepted species are:

Zinnia acerose: UT AZ NM TX Coahuila, Durango, Nuevo León, San Luis Potosí, Michoacán, Zacatecas, Sonora.

Zinnia Americana: Guerrero, Jalisco, Michoacán, México State, Nayarit, Veracruz, Oaxaca, Chiapas, Guatemala, Honduras, Nicaragua.

Zinnia angustifolia: Jalisco, Durango, Chihuahua, Sinaloa, San Luis Potosí.

Zinnia anomala: TX, Coahuila, Nuevo León.

Zinnia bicolour: Chihuahua, Durango, Jalisco, Guanajuato, Nayarit, Sinaloa.

Zinnia citrea: Chihuahua, Coahuila, San Luis Potosí.

Zinnia elegans from Jalisco to Paraguay; naturalized in parts of USA.

Zinnia flavicoma: Michoacán, Oaxaca, Guerrero, Jalisco.

Zinnia grandiflora: AZ NM TX OK KS CO Tamaulipas, Nuevo León, Coahuila, Chihuahua, Sonora.

Zinnia haageana: Guanajuato, Jalisco, Michoacán, México State, Oaxaca.

Zinnia juniperifolia: Tamaulipas, Nuevo León, Coahuila.

Zinnia maritima: Guerrero, Colima, Jalisco, Nayarit, Sinaloa.

Zinnia microglossa: Guanajuato, Jalisco.

Zinnia oligantha: Coahuila.

Zinnia palmeri: Colima, Jalisco.

Zinnia peruviana: widespread from Chihuahua to Paraguay including Galápagos and West Indies; naturalized in parts of China, South Africa, USA.

Zinnia purpusii: Chiapas, Guerrero, Colima, Jalisco, Puebla.

Zinnia tenuis: Chihuahua.

Zinnia venusta: Guerrero.

Zinnia violacea: Mexico, Central America, Colombia, Venezuela.

Cultivars developed from *Zinnia elegans* are well known of the twenty or so species in the Zinnia genus. Second well known genus *Zinnia angustifolia* was exploited for improvement. The chromosome number of *Zinnia angustifolia* is 2n=22, whereas in case of *Zinnia elegans*it is 2n=24. Some species of *Zinnia* are given below.

Zinnia acerosa: Commonly known as Desert zinnia or White zinnia. It is perennial. Sub shrub with slender, woolly stems and long narrow waves and flower heads white rays. It is 30 cm tall and 60 cm spread. It blooms in March-April. The Desert zinnia is useful in harsh environment where it receives minimum care.

*Zinnia angustifolia: Zinnia angustifolia*is annual, spreading, bushy, trailing and 30-45 cm tall. The leaves are 5-7 cm long. The foliage is narrow. The blooms are single or double, small daisy like blossoms in colours of white, orange, reddish brown, red, gold and yellow. Excellent for bedding, edging and containers.

Zinnia elegans: It is most popular and very useful species among the entire existing zinnia. It has wide variation in plant height (20-120cm), flower size (2-12 cm) and different flower colours. Flowers are single, semi double or double. Ray florets are flat, ruffled and quilled. Number of hybrids and new cultivars have been developed from this species and used in garden and as cut flower.

*Zinnia grandiflora: Zinnia grandiflora*is a subshrub having short leafy stems and numerous small flower heads as name grandiflora indicated. Blooms are nearly round, yellow, orange rays. Grown in South Arizona to Mexico and east Kansas.

Zinnia haageana: Commonly known as Mexican zinnia. It is an annual and requires warm days for good production. Most often flowers occur in either yellow or orange with an accent in red.

*Zinnia linearis: Zinnia linearis*is a perennial plant. Plant has a spreading habit and grows upto 25 cm height. It bears numerous flowers of white, yellow and orange colours. It is ideally suited for edging, beds, rock garden, hanging baskets and mass effect.

*Zinnia peruviana: Zinnia peruviana*was grown in the nineteenth century gardens and was sold by Philadelphia nurseryman BenardMc Mohan in 1804. This is 90-110 cm tall having yellow and orange colour flower. It bears flower throughout the summer.

Cultivars

Zinnia has many types, depending on the shape and size of flower, plant height and flower colour. There are a large number of cultivars. These are grouped in several types. The cultivars of several groups are given below:

Dahlia group: Blaze (orange scarlet), Canary Bird (yellow), Dream (lilac), Eskimo (creamy white), Meteor (dark red), Oriole (orange scarlet), Polar Bear (white), Yellow Marvel (yellow).

*Giant of California group:*Brightness (pink), Cherry Queen (rose), Orange Queen (golden orange), Lavender Gem (lavender), Purity (white).

Cactus group: Bonanza (golden orange), Cherry Time (cherry rose), Empress (pink), Fire Cracker (red), Princess (salmon pink), Red Man (scarlet), Snow Man (white), Sun Gold (light yellow), Sunny Boy (deep yellow).

Propagation

Seeds of various cultivars of zinnia are available easily and germinate quickly. Therefore, common zinnia is propagated through seeds. It germinates well when temperature is higher than in colder months, *i.e.*, December and January. For raising seedlings, seeds can be sown in raised nursery beds, pots or seed boxes. For large scale nursery, beds are thoroughly prepared by digging of soil. Well rotten farmyard manure or leaf mould of 3.0 kg/m2 to be incorporated in the beds. Beds should be drenched with carbendazim (0.2 per cent) atleast 48 hours before seed sowing. Sowing time of zinnia seeds under Indian conditions is February month. Seeds are sown thinly in lines and covered with mixture of sand and sieved leaf mould (1:1 ratio) followed by watering through watering can. For entire period nursery beds should be kept moist. Seed germinated in 5-7 days at 22°C temperature. Zinnias grow poorly and become chlorotic at temperature below 15°C. Ideal temperature for zinnia is 21°C.

Field Preparation

Field should be ploughed and cross ploughed till the fine tilth. For this stage, 3-4 ploughing is required. Zinnias require sunny situation, hence area should be exposed to full sun. During last ploughing farmyard manure at 4 kg/m2 is required for proper growth of plant and flowering. Drainage is most important for good stand of crop. It has also been observed that crop can be adversely affected if there is more moisture in the field and it also encourages diseases like powdery mildew and root rot. Therefore, all the plots should be well levelled and provision for drainage of excess water after irrigation. Plots should be prepared of convenient size and required dose of fertilizers should also be incorporated before transplanting of seedlings.

Planting

Seedlings having 3 to 4 true leaves are ready to transplant. It takes about 25-30 days from sowing. Uniformly grown seedlings are to be taken for transplanting. In different other countries zinnias are sown directly to the field and it requires more care at early stage. Production of crop is influenced by transplanting time. Seedlings transplanted during March resulted in maximum plant height, plant spread, number of branches and flower and seed yield than transplanting in April, May and June. Transplanting of seedlings in May and June also causes flower head rot when they are in peak flowering stage *i.e.*, rainy months due to high humidity. Hence, it is advisable to transplant zinnia during March for vigorous plant growth and higher flower production. Proper spacing is a very important aspect for zinnia and it may also vary greatly according to cultivar and time of transplanting. Narrow plant spacing is better for flowering as it reduces side branching. It is best for single harvest. Wider spacing at 60 cm between rows is better for multiple harvests. In bed culture, 15 cm between plants was found satisfactory for summer production of zinnia under Florida conditions. Plant spacing at 40×30 cm is found suitable for transplanting of zinnia under Indian conditions for good plant growth; flowering and seed yield per plant, whereas, closer spacing (30×20 cm) produced more seed yield per unit area. Zinnias are sensitive to root disturbance, therefore, care should be taken during transplanting. Transplanting should be done in evening hours and after transplanting plots should be irrigated immediately to avoid transplanting shock and better establishment of the crop.

Manures and Fertilizers

Before initiating a fertilizer programme, test of the soil for nutrient content is required. The increased water requirement for flower production creates and increases requirement for fertilisation. The application of fertilizers should coincide with crop needs and nutrient status of the soil. In zinnia, nitrogen deficiency causes reduction in growth and flower production. Phosphorus deficiency retards vegetative growth and delays flowering and deficiency of potassium results in reduction in number and quality of flowers. Application of 100 kg nitrogen, 100 kg phosphorus and 200 kg potassium along with 30 tonnes farmyard manure per hectare as basal dose and 100 kg Nitrogen/ha as top dressing at one month of transplanting increases flower and seed production in zinnia under Nadia conditions of West Bengal. The fertilizer application of 15 g/m2 of nitrogen, 20 g/m2 of each phosphorus and potassium as basal along with farmyard manure at 4 kg/m2 as basal dose and 15 g/m2 as top dressing was found beneficial under Ludhiana (Punjab) conditions. Application of 200 ppm Nitrogen atleast once a week is found beneficial for proper plant growth and flowering. Low level of boron may cause terminal bud blasting and slow branch development, whereas, high level of boron may delay flowering by 12-15 days. Boron application of 0.25 ppm produces healthy and vigorous plants. In general, zinnia plants being grown for cut flower requires higher amount of nitrogen than the same plants being grown for the display purpose.

Irrigation

Zinnia is a fast growing plant and makes vegetative growth rapidly in early stage and then enters to reproductive phase. In all the vegetative and during flowering and flower production, sufficient soil moisture is essential. The amount and frequency of water required will vary with the weather and maturity of the crop. Insufficient water will reduce crop production and quality, whereas, a consistently saturated soil will reduce growth and promote development of root rot. During early stage light and frequent irrigation is provided at a consistent rate so that it helps in the development of vegetative shoots and leaves. After the emergence of crop, the irrigation is provided at an interval of 2-3 days to promote flowering.

Harvesting

Harvesting of stems should be done during cool part of the day *i.e.,* morning or late hours of day when plants and flowers are free from dew and moisture. Full open flowers should be harvested. Use of clean, disinfected harvest containers and cutting utensils is essential. Stems are cut as long as possible but leaving enough nodes to ensure future production. Excess and damaged foliage from the stems should be stripped off. Newly harvested stems are placed in commercial floral preservative or citric acid (3.5 pH). Transportation of stems from the field as soon as possible to cold storage or a cool, shady place is required to enhance vase life of flower. Zinnia can be stored at 4°C for four days without affecting vase life. Generally vase life of zinnia is about 5-7 days.

Yield

Higher plant densities result in higher yield per unit area. In general, closer spacing produce more marketable stems per square meter than wider spacing. Seedlings transplanted during March produce more flowers, whereas, late transplanting reduces yield and flower quality because it bears flowers late transplanting reduces yield and flower quality because it bears flowers during rainy months. Flower yield may range from 10-12 tonnes/ha based upon cultivar and climatic conditions.

Postharvest Management

For fresh zinnias, leaves from the bottom half of the stem should be removed. Grading of cut flowers can be done according to length and flower size. Bunching can be done as per specifications. Recutting of stems under water to a uniform length is required. Bunched stems then placed in clean, disinfected containers in commercial floral preservative and stored at 3-4°C with 85-90 per cent relative humidity. Packaging and packing depends on specification of buyers. Generally flowers are delivered in buckets of floral preservatives for local markets. Flowers for regional, national and international markets are packed dry in boxes. Standard flower boxes are 50 cm wide and 25 cm deep with length ranging from 100-140 cm. they are often insulated. Most fresh speciality cut flowers are air freighted to market because of their short vase life and high value.

Drying of Flowers

Zinnias can be dried successfully and used in various ways. Demand of dry zinnia flowers is increasing tremendously. Available in a wide range of colours and sizes, zinnia provides a round flower shape and this unique feature is observed only in few dried flowers. Except for the deep reds and scarlet which usually turn an undesirable black-red, most coloured zinnia can be dried well. Removal of most of the stem should be done because the flowers dry most efficiently in shallow containers. Selection of drying substance such as white cornmeal, sand, borax, kitty litter, silica gel or a specially formulated product is important. A material that will spoil the flowers or be difficult to remove should be avoided. The flowers with stems are placed on a thin layer (1.2-1.8 cm) of drying substance in a container 7-10 cm deep. Careful pouring of the drying substance over around and through the petal to cover the flowers is an important operation. Quality of dry flower is influenced by drying conditions and temperatures. Sun drying causes more degradation of various pigments. Cabinet drying at 50°C also reduces pigment content of flowers. Room drying is found beneficial and there was less degradation of pigments like carotenoids and anthocyanin. The surround and cover drying method preserves the flower shape best. Hanging zinnias upside down can cause the petals to reflex inward and lose the desired round, daisy shape. Drying upright, zinnias often lack the strength to support the flowers as they dry. The succulent nature of the flower prohibits the use of glycerine for drying. Because most of stem is removed in the surround and cover method, many short stemmed and cultivars that are unsuitable for the fresh cut flower market can be used for drying.

Insect-Pests

Aphid

Zinnia is damaged by black bean aphid (*Aphis fabae*). It clusters on tender terminal shoots and leaves resulting in stunting and leaf curling. The aphids are easily controlled with malathion or metasystox at 0.2 per cent.

Leaf Hopper

Two types of hopper namely the red banded leaf hopper (*Graphocephalacoccinea sp.*) and the six spotted leaf hopper (*Macrostelesfascifrons*) feeds the zinnia plants. They are controlled by spraying of dimecron (1 ml/l) or rogor (2 ml/l) after appearance of the insects.

Bug

The four lined plant bug (*Poecilocapsuslineatus*), the long tailed mealy bug (*Psuedococcusadonidum*) and the bishop bug (*Lygusrugulipennis*) damages leaves, buds and flowers of zinnia. Malathion or Methoxychlor spray at 0.2 per cent is found effective to control most bugs.

Mites

Three types of mites, broad, cyclamen and two spotted affect plant growth and flowers of zinnia. They cause distorted foliage and shrivelled and discoloured bloom. They occur in hot season and damage leaves. Spray with metasystox or kelthane (2 ml/l water) from time to time is needed.

Nematodes

Foliar Nematode

Stunting, foliar browning, bud deformity are caused by *Aphlelncoidesritzemabosi* nematode. Soil testing for presence of foliar nematode is required. Application of Aldicarb is recommended as control measure.

Root Nematodes

Foliage yellowing, wilting, plant stunting, poor root development, root galls are caused by *Meloidogyne incognitai*, whereas dark root lesions are due to *Pratylenchus sp*. They may also cause angular spots on leaves. Phorate 10G should be applied in the soil to control these nematodes.

Diseases

Fungal Diseases

Leaf Spots and Blight

These diseases are caused by *Alternaria zinnia*. Symptoms are small reddish-brown leaf spots with gray centers. Dark brown cankers may develop on stems and flowers may be spotted or blighted. Proper plant spacing and good air circulation are necessary to minimize diseases. Application of Maneb or Mancozeb (0.2 per cent) at 7-10 day interval is found effective to control the diseases.

Botrytis Blight

Commonly *Botrytis cinerea* effects greenhouse plants. Large area of petals, leaves and stems turn brown. Affected plant parts develop a dusty gray covering during humid conditions. Proper plant spacing and good air circulation are necessary to control this disease. Various fungicides like Maneb, Mencozeb, Vinclozolin, Iprodione or Chlorothalonil were found effective. Mencozeb (0.2 per cent) at 7 to 10 days interval is a very effective control measure.

Other Leaf Spots

Several fungi (*Ascochyta* sp., *Cercospora zinnia, Cibrinia* sp. or *Ovulina* sp., *Corynespora* sp., *Curvularia* sp., *Entomosporium* sp., *or Septoria* sp., *Helminthosporium sp.* and *Volutella sp.*)may cause leaf spotting. Therefore, diagnostic techniques for proper identification of the pathogen are required. Leaf spotting fungi can be partially controlled by proper plant spacing and adequate air circulation. Several fungicides, including chlorothalonil, iprodione, maneb, mencozeb and thiophenatemethyl are effective in controlling certain leaf spot diseases.

Powdery Mildew

Zinnia is frequently infested by *Erysiphecichorecearum genus sp.* causing white, powdery growth on the leaves. It is more prevalent in the late summer and early fall. Preventive fungicide applications are necessary for adequate control of powdery mildew. Application of chlorothalonil, fenarimool, sulphur, triforine, triadimefon or thiophanate-methyl is effective to control this disease. Any one fungicide can be applied at 7 to 15 days interval as per requirement.

Root Rots

It is caused by *Pythium sp.* and *Phytophthora sp.* Symptoms are dark discolouration on roots and lower stem. In low, wet areas, plants may fail to emerge. Stunted plants turn yellow, wilt and die. Planting on raised bed to allow for good drainage is necessary. Pre plant dressings of ethazole or metalexyl helps to supress the damping off of roots in problem areas. Soil drenches or ethazole or metalexyl or foliar sprays with fosetyl-Al are effective control measures.

Fusarium and Rhizoctonia Rot

These diseases are caused by *Fusarium sp.* and *Rhizoctoniasolani.* Symptoms are dark discolouration on roots and lower stem and poor seedling emergence. Older plants may be stunted, wilt and die. Young plants damp off, stunted old plants turn yellow, wilt and die. Seeds should be free from seed borne diseases. Application of PCNB, iprodione or thiophanate-methyl as soil drench was found effective to control these diseases.

Wilt

Wilt is caused by *Fusarium oxysporum.* A brown discolouration can be found in the internal vascular tissue of stem. Leaves turn yellow progressively from the bottom of the plant to the top. Infected plants are stunted or killed. Location of planting beds should be changed in next crop. Seed treatment with bavistin or spraying of same chemical at 0.2 per cent can be applied at weekly interval to control this disease.

Bacterial Diseases

Bacterial Leaf Spot

Xanthomonas sp. is a causal organism of this leaf spot. Small water soaked spots with yellow halo. It is difficult to diagnose. Keeping plant surfaces dry by providing adequate circulation is necessary. Seed treatment with mercuric chloride or spraying of plant with streptomycin sulphate is recommended.

Viral Diseases

Beet curly top virus, cucumber mosaic virus, tobacco etch virus, tomato spotted wilt virus: Yellowing, plant stunting, mottling, vein clearing, necrotic etching, ring spots, shortened internodes are symptoms of viruses. There are no chemical controls once plants are infected. Infected plants should be rogued out immediately. Removal of weed plants (alternate host) is required to control the insect vector. Spraying of rogor at 1.5 ml/l at weekly interval is necessary.

Aster Yellows

Symptoms are stunting, shortened internodes yellowish green, deformed flower heads and foliage. Infected plants should be rouged out. Systemic insecticide like metasystox or rogor (0.15-0.20 per cent) should be applied at 7-10 days intervals.

18

Floral Ornaments and Flower Arrangements

Possibly floral garlands is one of the oldest method of using flowers for decoration. Even today the tradition of putting a floral garland round the neck of men and women on special functions such as marriage and welcome ceremony, is very much prevalent. Sweet scented flowers are preferred than others for making a garland. The most sought after flowers are Jasmine and tubrose, though other flowers such as marigold, chrysanthemum, Tubernaemontana cardnearia and rose are also used for garland making often a flower of lotus is used as the locket of the garland or in between the garland of other flowers. Red hibiscus garlands are offered to Goddess Kali and worn by a Tantrik who worships Kali. Garlands are made up of only one type of flowers or a combination of different flowers. The flowers are held together with the help of a cotton, nylon or a silk thread inserted in it with the help of a needle. For heavy garlands fine wire strings are used. Garlands often have a pendant or locket which is also made of flowers and is generally given a heart shape. Besides garlands, the other important floral ornaments are bangles and floral crowns. The bangles either made of Jasmine or tuberose among the fragrant flowers and T.M.C. and marigold among nonfragrant flowers. Floral crown are made of tuberose. Ear ring and Bajubandhs made of flowers are also used in ceremonial dances and other functions.

Hair Decoration

The temptation of women to decorate their hair with flowers is an age old custom. Here again, the fragrant flowers get preference over others in hair decoration. In south India flowers such as crossandra and Barleria are widely used for hair style (Gajra). The practice may be to put one or more flowers singly in the plaits of hair or to make minigarland like short chain of flowers and damage this from the hairdo or to wrap around the hairdo. The chain may be made out of a single fragrant species of flowers such as jasmine, tuberose or *Michelia champaka* or a mixture of flowers and decorative foliage. In south India flowers of Barleria and Crassandra are widely used for making the chains either singly or in combination with jasmine. Roses are used singly in decorating the hair. In Hawaii and in islands of Tahiti, flowers of cattleya orchids are commonly worn in the hair.

Veni

A special kind of flower arrangement is widely used in south India to decorate the long plait of hair (Veni) during some ceremonies such as marriage or at the time of Bharat Natyam dance recital. This flower arrangement consists of 90 cm long and 7-10 cm broad veni like structure which may be made of a frame of a comparatively hard cardboard or some leathery and tough leaves over which different flowers are arranged decoratively covering completely the background frame. This is attached to a hair style in which the tress of hairs are made into a plait by interlacing.

Rangoli

The drawing of pattern with the help of dry colour or a coloured paste on the floor is the common practice especially among the Hindus in India. Rangoli with flowers is also a common practice in India especially in the south. Generally the petals of different flowers are taken out and are arranged in various patterns. Intact flowers of small flowered chrysanthemum and others can also be used for this purpose.

Floral Bouquets

It is usual practice to present a nice floral bouquet on the occasion of birth days, marriages or to welcome important dignitaries. Bouquets may be of many types and shapes. The usual shapes are flat and round types but arranging bouquets in floral baskets is also considered as of high taste and beauty. For making a flat bouquet a hard poster paper in white or any other pleasing colours matching with the flowers is used having length of 45-75 cm. It is given a conical shape by folding the base or cutting the paper in that shape. Over this paper, aluminium foil white or coloured may be laid to make the bouquet colourful. The spikes are laid flat over this paper with the end having the flowers placed on the broader end and the base stems converging on the narrower side. The stems may be held in position with the help of cellotape stuck in 2-3 places against the paper backing. Sometimes ornamental foliage such as Thuja and Ashok are also spreadover the paper before arranging flower stems. Flowers such as gladiolus can be arranged in one plane since the numerous florets can cover the whole spread of the paper. The cut end of the flower stems at the narrow end of the bouquet should preferably be wrapped with wet moss and covered with aluminium foil. The stems at the base of the bouquet are tied tightly with thieve/ thread to prevent displacement. The twine is covered by wrapping around a nice silken ribbon which is patterned into a floral matif. A cord may be attached near the ribbon announcing the name of well wisher along with the message of goodwill. In round shape bouquet flowers are arranged in whorl which takes the shape of a cone, the stem end becoming tapering while the flower end comes in a round whorl. On the outer whorl some ornamental foliage is backed up to make the bouquet firm as well as decorative. The foliage used for this purpose are fronds of ferns, Thuja, Ashok, *Nandina domestica* and Crotons. The base of the bouquet is tied firmly with gunny twine which then camouflaged with a silk ribbon. The ribbon should be tied artistically. Bouquets in the form of flower and foliage arrangement, a nice basket is selected in which the foliage and flowers are arranged other insert materials such as coloured festoon papers (tissue paper), ribbons, Jari and drift wood may also be used for this type of bouquet. To keep the flowers fresh for longer periods is to put a small wet ball of cotton at the cut end of the stem and wrap the ball with a small piece of polythene, the open ends of which are secured with the flower stem by the help of a rubber band to prevent the moisture escaping out of the cotton. The whole arrangement may also be encased with a cellophane paper, if possible.

Buttonholes

This is another fascinating item in floral decoration. These are generally worn by males in their coat collars near the chest where a provision is made by criss-crossing two threads to hold the buttonhole but there is no harm if the ladies decorate their hairdo with a buttonhole besides they can bear in their coats. The most suitable flowers for this purpose are roses and orchids but other flowers such as carnation, camellia and daisy may also be used. This is prepared by first taking a small portion (about 5-10 cm long and 4-7 cm wide) of a leaf of a Thuja or *Asparagus plumosus* the most commonly used materials for this purpose and then fixing the desired flower or flower bud in position with the help of thread. Often a small coconut broomstick is attached at the base of a buttonhole to facilitate insertion in the appropriate place. Jari threads may be used for brightening up the look. To camouflage the broomstick aluminium foil may be wrapped around. The Japanese form of flower arrangement is known as Ikebana. The western style lays emphasis on mass arrangement while Japanese form uses only a few blooms and branches and has a spiritual and religious background.

Wreaths

Funeral design has two type. Cross constructed and Wreath

Flower Shows

The primary objective of flower show is to create interest among the general public to grow quality flowers and maintain beautiful gardens around their house. The display of quality exhibits inculcates the spirit of healthy competition among the participants. Flower shows give an opportunity to the people to know the wide range of plants that can be grown in the locality. The other advantage is that a visitor gets a chance to see all the best materials at a time in one place. A flower show should be a place for discuss the various garden problems and to find out the ways for solving each others difficulties. Each experience with others but unfortunately due to professional jealousy people sometimes do not want to divulge their secret to success, lest others copy and score over them. It is also an occasion to demonstrate how a perfect exhibit can be grown. Besides all these the aesthetic value of a show cannot be overlooked.

The success of a show depends much upon the co-operation between the exhibitor and the member of flower show committee. The judging should be above criticism though one hundred per cent success can not be achieved on this last score, as controversies are often bound to crop up even with the best judgement. Some preliminary spadework has to be done to make a show successful especially if it is a *Maiden* one. Firstly the interested garden lovers of the locality should get together, elect a president, honorary secretary and form some committees to run the show.

(1) A general or a ground committee – Preparation of site etc.

(2) A editorial committee – Schedule/rules and regulation, printing tickets etc.

(3) Finance committee – Financial aspect

(4) Plant committee- Different plant sections

(5) Floral committee – Cut flowers and floral arrangements

Flower vases, bamboo stakes, show passes, benches and tables for exhibits should be made available to the participants. Arrangements should be made to open refreshment stalls. Nurserymen, Seedmen, Companies selling agricultural implements, chemicals etc., should be allowed to open their stalls on rental basis. Different educative charts on horticulture should be displayed. A film show on flowers and gardens could be arranged in the evening accompanied by a supporting talk.

Tips to Exhibitors

The first and foremost thing is to get a schedule of flower show well in advance and to convince oneself on the requirements. For example roses are grouped in several types: as hybrid tea, floribunda etc., and one has to exhibit right type of rose in the appropriate class. After going through the schedule has to be decided in which groups, the entries are to be made. Once the decision is made plants are raised accordingly. The saving is to be staggered at intervals of 4-7 days so as to avoid disappointment as a result of casuality when planted in one lot and to ensure that atleast one group of plants is in perfect condition during the show time as it may so happen that one group from a particular sowing date fail to open the flowers on the scheduled date for climatic or other reasons. Plants for exhibition needs extra feeding from the date flower buds start to appear with liquid manure. To obtain large flowers all auxillary buds in flowers such as carnation, marigold, dahlia etc. should be disbudded as soon as they appear leaving only apical bud to bloom. The seasonal flowers should be grown in the appropriate-sized pots. To make plants bushy pinching should start at an early date for flowers such as brachycome, carnation, marigold, zinnia etc., and the operation repeated frequently. Pots are to be cleaned properly or lightly painted with terracotta red. No plants should be displayed in a crowded fashion.

Precautions

1. A few extra plants are to be taken to the show than required as plants may get damaged while in transit.
2. Staking and covering of flowers with tissue paper should be done as a precaution against being bruished.

Staging the Exhibits

The staging of exhibits is some sort of an art and the value of the artistic display cannot be minimized in winning prizes. A stage rehearsal of grouping of plants and flowers may be undertaken at home to get the best combination and placement. All plants are displayed distinctly to make the task easy for the judges. The colour arrangement should be artistic and harmonize with each other at the background. Dead leaves and dried flowers should be removed carefully without leaving a scar. The leaves should be washed with clean water. Artificial polishing of leaves is forebidden. The exhibits are to be correctly lebelled with indelible ink.

Cut Blooms

The blooms for exhibition in the cut flower section are raised with extra feeding with liquid manure and disbudding. In temperate country, a rose bloom is cut in the previous afternoon but under our condition, the bloom is cut in early morning of the day of the show. A rose bloom should atleast be half open and not more than ¾th open. It may be padded up with the leaves to keep the blooms upright in the vase. The stem should be cut below the vase level.

Judging

Only competent persons having sound knowledge on the subject should be appointed judges. The integrity of judges should be above suspicion. In a small show a group of 3 judges will be enough to judge all the groups. But in larger size show, a group of 2-3 judges are assigned for each group. The duties of a judge are manifold, they should penalize a crowded display. Plants showing signs of wilting which are lifted from the ground and potted should be disqualified outright, plants grown in over or undersized pots should be penalized. Plants not grown in the locality should not be entertained. An artificial colouring of foliage or bloom should not be permitted. Biennials or Perennials impatiens, carnation etc. which flower naturally in one year may be allowed in the annual section but dahlias and carnation raised from cuttings should not be allowed as annuals. Specimens not labeled or incorrectly labeled should be rejected. Only one plant per pot should be allowed. A group should contain only the specified number of plants and varieties mentioned in the schedule.

Score Cards

There is a set pattern for judging different categories of plants and blooms.

(A) Foliage Plants

Plants are judged on the basis of variety of specimen, culture, appearance or health of the plants. Points may be awarded for display also. The marks may be distributed as follows: culture 4, quality, including rarity 3, Appearance of foliage 2, display 1.

(B) Annuals

- ☆ Quality and quantity of bloom – 4
- ☆ General appearance – 3
- ☆ Rarity of the specimen – 2
- ☆ Display and colour scheme – 1

(C) Cut Flowers

- ☆ Size and form – 5
- ☆ Colour and Rarity – 3
- ☆ Arrangement – 2

Rose society of India has divided the marks for judging specimen rose blooms in vase or bowl as follows:

Colour 20,	Substance (Texture) - 20	Stem – 10
Form 20,	(firmness and thickness of petals)	(Foliage and stem carry no
Size 20,	Foliage - 10	marks when in boxes)

Floral Display

Judged by aesthetic and artistic value

- ☆ General setup – 5
- ☆ Colour scheme – 3
- ☆ Novel ideas – 2

Bouquet

- ☆ Size and artistic value – 5
- ☆ Arrangement – 3
- ☆ Novelty – 2

These are two type of bouquet, namely Posy and shower and each should be placed in a separate groups while judging.

Buttonholes, Sprays, Twig, Garlands etc.

- ✩ Arrangement – 5
- ✩ Colour or appeal and freshness – 3
- ✩ Novelty – 2

Garden Competitions

Along with the flower show, many horticultural societies run garden competitions, generally a week ahead of the actual show. All amateurs are invited well in advance to enter themselves for the competition and are advised to register their names 3-4 months in advance of judging dates. The gardens should be placed in different categories such as private residential gardens, Factory gardens, Institutional or government gardens etc., and judged separately. The aim of such competition is to make the people garden minded and correct the deficiencies by constructive criticism. Generally the amateurs tries to pack the garden with more colour by planting masses of cannal flowers but a garden without a proper landscaping effect is no garden at all. A garden should have an "All the yearround garden" an amateur may crowd this garden with many features, which may look out of place. For instance, sundial in the lathhouse where sun never enters or a pergola leading to nowhere or a Japanese lentern in a formal garden etc., may look odd and spoil the naturalness of the garden. Similarly, a large shade tree such as the rain tree will be completely out of place in a small home garden.

Some features in a garden are definitely needed but this should be created in such a manner as to enhance the beauty of the garden and not added just as a personal pride of the owner. The lawn should be kept trimmed and fed with liquid manure to make it lush green. The borders of the beds should be cut neatly and the garden kept tidy by removing dead wood, leaves, grasses and other rubbish. The judge will visit all the corners of the garden and any artificial camouflaging such as putting potted plant in an empty flower bed, will be detected and penalized. It is rather advisable to keep empty beds neat and clean. Any specialties in a garden like rare collections of palm, cacti and succulents and bonsai will receive special attention from the judges.

	Score Card	*Small Garden*	*Large Garden*
1.	**Overall effect of the garden**		
(i)	General landscape and planning	75	100
(ii)	Shrubbery	30	30
(iii)	Proper use of climbers	20	20
(iv)	An all the year round effect (less annuals)	30	30
(v)	Features such as statue, bird bath etc.	30	30
(vi)	Specialities of a garden *e.g.*, bonsai, palms, cacti etc.	40	40
2.	**Lawns and grasses in other places** (Tennis court, round the annual garden, crazy paving) (quality and colour of grass, mowed or not, weed-free or not)	75	100

	Score Card	*Small Garden*	*Large Garden*
3.	**Flowering annuals**		
(i)	Uncommon cultivars and varieties	10	10
(ii)	Design of the beds, site selection, suitability of place, harmony in colour selection	50	50
(iii)	Quality and Quantity of bloom	40	40
4.	**Maintenance and attention**	50	50
	Total	**450**	**500**

19

Pest Management

Cultivation of Ornamental Crops on commercial scale is fast gaining importance in India. These crops are being cultivated for their importance as cut flowers both for domestic and export purpose. However, the large scale production of these crops is hampered by several limiting factors. Damage caused by insect-pests is one of the important factor limiting quality and quantity production of these crops.

Commercial cultivation of ornamental is being taken up in India for their importance in cut flower industry, manufacturing drugs and manufacturing perfumes, respectively. The large scale production of these crops is hampered by several limiting factors. Insect-pests are one of the important limiting factor for quality and quantity production of these crops because of damage they cause. Table 19.1 gives information on some of the important insect-pests, attacking these crops, their damage and control.

Table 19.1: Symptoms of Damage and Control of Various Insects of Various Crops

Sl.No.	*Crop*	*Insect-pests*	*Nature and Symptoms of Damage*	*Control/Management*
1.	Rose	(1) Red Scale (*Aonidiella aurantii*)	Reddish-brown encrustations are seen on lower portions of old stems and growing shoots. Damage by sucking plant sap. Affected branches dry off.	Removal and destruction of infested branches. Application of pongamia oil soon after pruning to the affected part. Applications of carbofuran granules @ 1.5 kg ai/ha.
		(2) Thrips (*Rhipipho thrips*, *Thrips* sp.)	Both nymphs and adults such cell sap from tender leaves and buds. Mottled and deformed young leaves with silvery patches and deformed flowers are the symptoms of damage.	Spraying of Metasystox (0.5 per cent) or Dimethoate (0.05 per cent) or Methyl parathion (0.25 per cent).
		(3) Aphids (***Macrosophum*** sp.)	Found clustered on young shoots, buds and flowers. Both nymphs and adults suck sap and devitalize infested plant parts.	Phosphamidon (0.02 per cent) or Dimethoate 0.03 per cent. Sprays at 15 days interval provides effective control.
		(4) Caterpillar pests (***Helicoverpa armigera, Euproctis fratarna***)	Causes severe damage to leaves, flower buds and flowers.	Spraying of Endosulfan (0.07 per cent) or fenetrothion (0.05 per cent).
		(5) Beetles (***Adobetus*** sp., ***Apogamia*** sp.)	Grubs feed on roots.Adults active in nights and damage leaves by making irregular holes and punctures.	Treatment of soil with 5 per cent dusts of Aldrin or chlordane or BHC
		(6) Weevils(***Myllocerus***)	Causes severe damage to foliage by making regular cuts on leaves.	Spraying of Endosulfan (0.07 per cent) or Fenetrothion (0.05 per cent).
2.	Chrysanthe-mum	(1) Leaf minor (***Liriomyza trifoli***)	Larvae makes tunnels in leaf and feeds on it. Affected leaves dry in severe cases.	Use of yellow stick traps. Growing field bean as trap crop. Spraying of Diazinon 0.05 per cent orTriazophos 0.04 per cent.
		(2) Aphids (***Macrosiphoniella sanborni***)	Nymphs and adults suck sap from growing parts. Affected parts turn yellow and wilt.	Controlled effectively by spraying Nuvacron or Phosphamidon (0.05 per cent).

Contd...

Table 19.1–*Contd...*

Sl.No.	*Crop*	*Insect-pests*	*Nature and Symptoms of Damage*	*Control/Management*
		(3) Thrips (***Haplothrips*** sp. and ***Thrips tabaci***)	Feed on tender parts. Mottling and browning of leaves and flower buds are symptoms of attack.	Spraying of Nuvacron or Dimethoate (0.05 per cent).
		(4) Caterpillar (***Hedylopta indicate***)	Larvae rolls the leaves and buds together and feeds inside.	Collection and destruction of affected plant parts. Spraying of endosulfan (0.07 per cent) or Quinalphos (0.05 per cent).
3.	Gladiolus	(1) Thrips (***Thrips simplex***) (***Hedylopta indicate***)	Feeds on tender leaves and buds of growing spikes. Attacked spike becomes brown and deformed.	Controlled by spraying of dimeothate (0.05 per cent) or Metasystox (0.5 per cent).
		(2) Cut worms (***Agrotis pegotum***)	Larvae feeds on young growing plants and cuts at ground level.	Spraying of 0.04 per cent quinolphos.
		(3) Leaf feeding catepillar (***Spodoptera litura***)	Present at lower ride of leaves and skeletonizes them. Browning and drying of attacked leaves.	Collection and destruction of egg masses and young larvae. Spraying of Methyl parathion 0.05 per cent specially to lower sides leaves.
		(4) Moaly bugsthrips and mites on corms.	These 3 insect-pests are problem during storage. They suck sap and affected croms are deformed and ultimately dry.	Treatment of croms with Metasystox or Acephate or one clocks multiplication of these pests.
4.	Marigold	(1) Leaf hoppers (***Empoasca*** sp.)	Both adults and nymphs suck sap from leaves. Attacked leaves curl, turn yellow and fall off.	Phosphamidon at 0.05 per cent given control.
		(2) Bud/Head borer (***Helicoverpa armigera***)	Feeds on growing buds and flowers. Results in loss of flowers.	Spraying of Endosulfan 0.07 per cent controls pest effectively.
		(3) Mites and thrips	Suck sap from leaves and flowers. Infested flowers turn brown and dry in severe cases.	Spraying of Lethane (0.05 per cent) for mites and Rogor/Metasystox for thrips.

Contd...

Table 19.1–*Contd...*

Sl.No.	*Crop*	*Insect-pests*	*Nature and Symptoms of Damage*	*Control/Management*
5.	China Aster	(1) Pumpkin beetle	Causes extending damage to leaves and growing parts of young plants.	Spraying of quinolphos (0.7 per cent) or Nuvecron (0.05 per cent).
		(2) Caterpillars (***Phycita*** sp., ***Helicoverpa armigera***)	Larvae feeds on young buds and flowers causing heavy flower loss.	Endosulfan (0.07 per cent) or Nuvacron (0.05 per cent) controls the caterpillar pests. Effect light traps in fields.
		(3) Stem borer (***Platyptilia molopias***)	Larvae soon after hatching bores into growing stem and side branches. Affected stem become hallow and dry off.	Methyl parathion or Nuvacron spray at 15 days interval controls them.
6.	Tuberose	(1) Thrips (***Thrips*** sp.)	Both young and adult thrips suck the sap from leaflets and buds of growing stalks. Infested spikes get destertd and dry.	Controlled by Rogor or Metasystox.
		(2) Bud borer (***Helicoverpa armigera***)	Bores into young buds causes flower loss.	Endosulfan (0.07 per cent) or Fonotrothi (0.05 per cent) controls bud borer.
		(3) Aphids	They suck sap from buds and growing of spike which results in drying of flowers.	Spraying of Malathion (0.1 per cent) or Dimethoate (0.05 per cent).
7.	Carnation	(1) Carnation moth (***Tortrix pronubata***)	Larvae webs leaves with silken threads and feeds inside. Grown up larvae bores in to shoots and flowers.	Use of light traps spraying Triazphos 0.2 per cent controls pest.
		(2) Red spider mite (***Tetranychus utricae***)	Suck sap from lower side of flowers and leaves. Affected leaves look pale yellow dusty and dry ultimately.	Methyl demton (0.05 per cent) spray contimites.
8.	Orchids	(1) Mealy bugs (***Pseudococcus*** sp.)	Suck sap from leaves and flowers and secrete honey dew. Favours sooty mould development.	Collection and destruction of masses. Spraying of Methyl parathion.
		(2) Thrips (***Anaphothrips orchidaceus***)	Suck sap from buds and flowers and causes distortion, browing and drying of affected parts.	Metasystox or Nuvacron spray provides effective control.
		(3) Scales – 2 types Armoured and soft	Suck sap and destroy the affected plant parts.	Albicorb or Methyl parathion controls both scales.
		(4) Snails and slugs (***Deroceras leave, Achatina fulica***)	Damage roots, young leaves flower buds and opened flowers.	Collection and destruction is effective. Spraying of Metheldehydo (1 per cent).

Contd...

Table 19.1–*Contd...*

Sl.No.	Crop	Insect-pests	Nature and Symptoms of Damage	Control/Management
9.	Crossandra	(1) Scales two types (***Saissetia migra, Orthesia insigna***)	Cover complete plant and suck the plant sap. Affected plant loses vigour and dry off.	Removal and destruction of affective plants parts. Application of carbofuran G.R.-3 g/plant.
		(2) Spike borers (***Cacoecia epicyrta, Helicoverpa armigera***)	Larvae feed on developing spikes and prevent flowering.	Endosulfan (0.07 per cent) controls the pests.
10.	Jasmine	(1) Bud worm (***Hendecasisdu plifascialis***)	Larvae bores into immature buds and leaves fecal matter on affected parts.	Collection and destruction of outer bad parts. Quinolphos or diflubensonon provides effective control.
		(2) Gallery worm (***Elasmopalpus jasminophagus***)	Makes a allerus out of leaves, buds and tender shoots and feeds inside.	Nuvacron or Fodvalorate sprays controls the part collection and destruction.
		(3) Web worm (***Mausinoe*** sp.)	Larvae web leaves with silken threads and feeds on thereby skeletonizing.	Spraying of methylparathion and collection and destruction of infested parts.
		(4) Euryophid mite (***Aceria jasmine***)	Produces galls by feeding on leaves and tender shoots. Results into malformation and drying of affected parts	Traizophos or Nuvacron sprays give effective control of mite.
11.	Foliage plants			
	1. Coleus	(i) Lantana bug (***Orthesia insignis***)	Most common posts on coleus both adults and immature stages suck the sap and reduce the vigour of plant.	Spray Methyl parathion or Dimethonte at 0.04 per cent or chlorpyrephos at 0.05 per cent.
		(ii) Scales (***Larga aegyptiaca***)	Affected plant portions get distorted and dry.	Collection and destruction of infestel leaves.
		(iii) Mealy bug (***Ferrisia virgata***)		

Contd...

Table 19.1–*Contd...*

Sl.No.	*Crop*	*Insect-pests*	*Nature and Symptoms of Damage*	*Control/Management*
2.	***Codiaeun*** spp. (Crotons)	(i) Mealy bug (***Ferrisia virgata***)	Commonly occurring insect-pests on various crotons.	Spray any of the insecticides mentioned above for scales, mealy bugs and thrips.
		(ii) Scales (***Parasaissetia nigra***)	Both adults and nymphs suck the sap.	
		(iii) Thrips (***Heliothrips haenorrhoidalis***)	Attacked plant parts get discoloured, distorted and dry in severe cases.	
		(iv) Mites (***Tetranychus cinnabarinus***)		Use Kalthane (2.2 ml/l) or any other Muticide for control of mites.

Glossary

Abaxial: The side of an organ turned away from the axis, for example, the under side of a leaf.

Aberrant: Different from the normal.

Abnormal: Different from the usual or prevailing form or structure.

Abortion: Failure of development or imperfect development.

Abortive: Imperfectly developed or underdeveloped.

Abruptly pinnate: Pinnate but without terminal leaflet.

Abscission: The shedding of some part or structure as a leaf or flower at its base that break apart readily.

Acaulescent: Apparently without a stem that is with the main stem underground and only basal leaves and slender, leafless flowering stems appearing above ground level.

Accessory: Additional beyond the normal; extra.

Accrescent: Increasing in size after flowering, enlarging with age.

Accumbent: Applied to cotyledons with their edges turned towards the main axis of the embryo.

Aceriform: Similar to the leaf of a maple.

Acerose: Needlelike.

Achene: A dry one seeded fruit with a firm close fitting wall which does not open by any regular dehiscence.

Achene beak: The persistent and hardened style of an achene.

Achlamydeous: Having neither calyx or corolla.

Acicular: Needle shaped and very slender.

Acorn: The leathery fruit of an oak, containing a single large seed and enclosed basally in a cup formed from bracts.

Actinomorphic: Radially symmetrical, that is, capable of division by straight boundaries into three or more similar sections.

Active bud: A growing bud.

Aculeate: Covered with prickles.

Acuminate: With a long, tapering point set off rather abruptly from the main body.

Acute: With a pointed end forming an acute angle, that is, less than a right angle.

Adaxial: On the side towards the axis, as for example,the upper side of a leaf.

Adherent: Grown fast to a structure of a different bind, as for example, a stamen grown fast to a petal.

Adnate: Adjective of Adnation.

Adnation: Fusion with or attachment to another structure from the beginning of development.

Adventitious: A structure occurring in an unusual position as, for example, an adventitious root formed from a stem or a leaf.

Adventitious bud: A bud produced in a unusual or unexpected place, as for example near the point of injury of a stem or on leaf or a root.

Adventitious roots: Are those shoots that arise from the aerial plant parts, underground stem or from old roots.

Adventitious shoots: Are those shoots that develop on roots or inter-nodally on stems.

Adventive: Said of an introduced plant beginning to spread into a new locality or region.

Aerial root: Roots exposed to the air, as those of tropical plants growing on the trees.

Aestival: Of the summer.

Aestivation: Arrangement of the young flower parts in the bud.

Agglomerate: Crowded into a dense cluster but not joined synonymus Aggregate.

Aitinomic: As referred to parthenocarpy, the ability of a part to develop parthenocarpic fruits only in response to some stimulus external to ovary.

Alate: With a wing.

Albumen: A deposit of reserve food material accompanying the embryo. Such reserves often are in the endosperm.

Alga: A seaweed or one of the essentially microscopic pond scums. Plural algae.

Alliaceous: Having the odor or taste of garlic or onions.

Alternate: With a single structure of each bind occurring at each level of the axis appearing to alternate on opposite sides of the axis, but actually in a spiral arrangement.

Alveolate: With angular depressions forming a pattern like honey comb.

Alveolus: A depression with in an alveolate pattern.

Ament: A soft, usually scaly spike of small, apetalous, unisexual flowers, this usually falling as a single unit.

Amentiferous: With the flowers in aments *i.e.* catkins.

Amorphous: With no definite form.

Amphitropous: Describing an ovule bent back along and adnate to the funiculus, but with the micropyle not bent all the way back to the funiculus.

Amplexicaul: Clasping the stem, that is, the base nearly surrounding it.

Ampliate: enlarged.

Anastomosing: Joining each other and forming a network, the term being applied to veins as the veins of the leaf.

Anatropous: Descriptive of an ovule in which the body is bent backward along the funiculus and adnate to it.

Ancipital: Two edged.

Androecium: The male reproductive organs; in the flowering plants the stamens- a collective term for all those within a flower.

Androgynous: A term applied to an inflorescence which includes both staminate and pistillate flowers, the staminate ones being apical, the pistillate basal.

Androphore: A stalk or other supporting structure under stamens raised above their normal position in the flower.

Anemorphilous: With the pollen carried by wind.

Annual: A plant completing its life cycle in a year or less.

Annular: Ring like.

Annular thickening: Thickening of the wall of a xylem cell laid down in the form of a ring.

Annulus: A ring. In the ferns a crest like or other special structure on the sporangium.

Anther: The portion of the stamen which produces pollen.

Anther tube: A tube formed by coalescent anthers, as in the sunflower family.

Antheridium: A structure producing antherozoids, that is male gametes or sex cells.

Antherozoid: A male gamete (sex) cell of a plant.

Anthesis: The time when the flower expands and opens or the process of expansion and opening.

Anthocyanin: Any of a common class of pigments having colours ranging from labender to purple. These pigments are affected by the acidity or alkalinity of the cell sap, and they change colour in approximately the same way as litmus paper, tending towards red in an acid medium and blue in a basic medium. These pigments are water soluble and not associated with plastids.

Antipodal cells: Three cells at the end of the flowering plant megagametophyte opposite the egg.

Antrorse: Directed upward or forward.

Apetalous: Without petals.

Aphyllous: Without leaves.

Apical: Of the apex.

Apical placenta: A placenta at the distal (apical) end of the ovary.

Apiculate: Terminated by an abrupt, short, flexible point.

Apiculation: An abrupt, short, flexible point.

Apocarpous: With the carpels separate.

Apogamous: Developed without the joining of gametes, that is asexually.

Apomixes: In some species the embryo is not produced as a result of meiosis and fertilization, but form a cell in the embryo sac or surrounding nucellus which does not undergo meiosis but develops to form a zygote of same genetic makeup as the female parent.

Apophysis: A swelling on the surface of an organ as, for example, the somewhat swollen exposed portion of the cone scale of a gymnosperm.

Appendiculate: With a basal appendage.

Appressed: Lying tightly against another (Usually larger) organ.

Apterous: Without wings.

Aquatic: Growing in water.

Arachnoid: With slender, tangled hairs resembling the threads of a spider web.

Arboreous: Of tree like form, that is with a main woody trunk.

Arborescent: Of large size and more or less tree like but without the clear distinction of a single trunk.

Archegonium: A flask shaped organ containing an egg, that is, female gamete or sex cell.

Arctic: The area beyond timber line at high latitudes.

Arcuate: Forming a moderate curve or arc.

Areolate: Divided into small, marked-off spaces (Reticulate).

Areole: Diminutive of area; a small, clearly marked space. The term is used most frequently in reference to the small, special spine-bearing areas on the stem of a cactus.

Aril: A large appendage of the funiculus at the hilum (attachment area) of a seed. It tends to envelope the seed.

Arista: A stiff bristle.

Aristate: with a stiff bristle.

Articulate: with conspicuous segments or joints.

Artificial system: A system made up according to predetermined supposedly important characters.

Ascending: Arising at an oblique angle.

Aseptic: Absence of fungi, bacteria, viruses, mycoplasmas, or other microorganisms in cultures.

Asperous: Rough.

Assurgent: ascending.

Altenuate: With a long, tapering point, this usually set off rather abruptly from the main body of the object (ex. a leaf blade).

Auricle: An appendage shaped like the lobe of a human ear.

Auriculate: Having an artricle.

Austral: Southern.

Autogamy: When a flower is fertilized by its own pollen.

Autonomic: As referred to parthenocarpy, the ability to set fruits without the stimulus resulting from pollination.

Awl-shaped: With a narrow flattened body tapering very gradually upward into a point (subulate).

Awn: A long stout or stiff bristle.

Awned: With an awn.

Axenic: Totally free from association with other organisms.

Axil: The adaxial angle between two organs, particularly the angle between the upper side of a leaf and the stem.

Axile: On the axis, as for example, said of placentae at or near the centre of an ovary.

Axillary: In the axil.

Axillary bud: A bud in a leaf axil.

Axillary flower: A flower in the axil of a leaf or a bract.

Auxin: Auxin is a generic term for compounds characterized by their capacity to induce elongation in shoot cells. They resemble indole-3-acetic acid in physiological action. Auxin may, and generally do, affect other processes besides elongation, but elongation is considered critical. Auxins are generally acids with an unsaturated cyclic nucleus or their derivatives.

Baccate: Berry like, that is, flushy or pulply.

Banner: The upper and usually largest petal in the papilionaceous corolla of a plant of the pea family. Known also as a standard or a vexillum.

Barbed: With a rigid barb like the barb of a fish hook.

Barbellate: Diminutive of barbed (Still smaller, barbellulate).

Barbulate: With a beared of fine hairs.

Bars of sanio: Special thickening on the walls of some xylem cells.

Basal leaves: Leaves at the base of an herbaceous plant arising from several nodes separated by exceedingly short internodes occurring at about ground level.

Basal placenta: A placenta at the basal end of the ovary.

Basifixed: Attached at the base.

Beaked: Ending in a firm elongated slender structure.

Beared: With long or stiff hairs.

Berry: A fleshy or pulpy fruit with more than one seed and formed from either a superior or an inferior ovary. The seeds are embedded in pulpy tissue.

Bidentate: With two teeth.

Biennial: Completing the life cycle in two years. Biennial plants usually produce only basal leaves above the ground the first year and both basal leaves and flowering stems the second.

Bifid: Forked, that is, ending in two parts.

Bilabiate: Two lipped. A bilabiate corolla has petals in two sets commonly with two in the upper and three in the lower.

Bilateral: Two sided.

Bilaterally symmetrical: Capable of division into only two similar sections, which are mirror images of each other.

Bilocular: With two cavities, as an ovary with two seed chambers.

Bipinnate: Pinnate and the primary leaflets again pinnate.

Bipinnatified: Pinnatified with the primary divisions again pinnatified.

Bisexual: With both sexes represented in the same individual or organ *e.g.* with stamens and pistils in the same flower.

Blade: The broad, usually flat part of a leaf.

Bloom: A usually waxy; whitish or bluish powder covering the surface of a leaf, stem, fruit or other organ.

Blossom end: The end of an inferior ovary and the surrounding floral cup which supports the flower parts or at fruiting time the calyx and remains of other flower parts.

Boreal: Northern.

Boss: A protrusion, as for example, a protuberance near the centre of the apophysis of a scale of a gymnosperm cone.

Bract: A leaf subtending a reproductive structure, such as a flower or a cluster of flowers or an ovuliferous scale. Usually the leaf is specialized and at least somewhat dissimilar to the foliage leaves.

Bracteate: With bracts.

Bracteolate: With bractlets.

Bracteole: Diminutive of bract, a mere scale.

Bractlet: A minor bract.

Branchlet: A small branch, that is one of the smallest degree.

Branch primordium: The lump of tissue which develops into a young branch.

Bristle: A stiff hair.

Bryophyta: The mosses, liverworts and anthocerotae.

Bud: A growing structure at the Tip of the stem or a branch, with the enclosing scale leaves or immature leaves; a young flower which has not yet opened. There are vegetative buds and flower buds, as described in the definition above.

Bulb: An underground bud covered by fleshy scales, the coating formed from the bases of leaves.

Bulbiferous: With bulbs.

Bulbils: A small bulb. The word is applied to bulb like structures produced on the stems, usually in the axils of leaves or sometimes in the places where flowers ordinarily occur.

Bulblets: A small underground bulb.

Bulbous: Bulblike.

Bullate: Blistered or puckered.

Caducous: Falling very early. In the poppy family the caduceus sepals fall away when the flower open.

Caespitose: Growing in tufts or mats.

Calcarate: With a spur.

Callosity: A thickened hard structure.

Callous: With the texture of a callus.

Callus: Production of new unspecializing cell from a part of the plant, growing as a result of injury or wounding.

Calyculate: With small bracts in a series below the flower, these resembling a calyx.

Calyptra: A cap, cover or lid.

Calyx: A cup; the sepals of a flower, the outermost series of flower parts. Each sepal has the same number of vascular traces as a leaf of the species. If there is doubt it may be distinguished by this character cf. petal.

Calyx tube: A tube formed from the lower portions of the sepals. The term has been used loosely for (a) a floral cup or floral tube regardless of its origin, (b) a floral cup or tube formed by coalescence and adnation of the bases of the sepals, petals and stamens, (c) a perianth tube of the type in the monocotyledons formed by edge to edge adnation of the adjacent sepals and petals together with the bases of the stamens (d) and the portion of a floral cup or tube above the area of adnation to an inferior ovary.

Cambium: A group of cells which possess the power of active growth and division, and which is responsible for the production of old tissues in matured organs.

Campanulate: Bell shaped, that is, the shape of an inverted church bell ie. rounded at the attachment and with a broad flaring rim.

Campylotropous: Descriptive of an ovule which curves in such a way that the micropyle is near the funiculus or stalk but the side of the ovule is not adnate to the funiculus.

Canaliculated: With lengthwise furrows or channels.

Cancellate: Appearing to be a lattice work.

Canescent: Grayish- white or hoary, densely covered with white or gray fine hairs, these usually short.

Capillary: Like an elongated delicate hair or thread.

Capitate: In a dense cluster or head.

Capitulate: Diminutive of capitate.

Capitulum: A small head.

Caprifig: The wild or 'male' fig, the uncultivated form.

Capsule: A dry, many seeded fruit made up of more than one carpel and splitting open length wise at maturity.

Carina: A keel, literally like the keel of a boat. Sometimes said of a structure which simply is folded and creased and without a projecting keel on the crease.

Carinate: With a keel.

Carpel: A specialized leaf which forms either all or part of a pistil.

Cartilaginous: With the texture of cartilage that is, tough and firm but somewhat flexible.

Caruncle: An appendage at the attachment point of a seed.

Carunculate: With a caruncle.

Caryopsis: The fruit of a grass that is, a one seeded, indehiscent fruit with the pericarp adnate to the seed.

Castaneous: Chestnut coloured.

Catkin: A soft, usually scaly spike or raceme of small apetalous unisexual flowers, the inflorescence usually falling as a single unit.

Caudate: With a talelike structures.

Caudex: A largely underground stem base which persists from year to year and each season produces leaves and flowering stem of short duration.

Caulescent: Having a well developed stem above ground as opposed to having only what appears to be the stalk of the individual flower or cluster of flowers.

Cauline: Of or on the stem.

Cell: The structural and functional unit of living organisms.

Cell culture: This term is used to denote the growing of cells in vitro including the culture of single cells. In cell cultures, the cell are no longer organized into tissues.

Cell hybridization: The fusion of two or more dissimilar cells leading to the formation of a synkaryon.

Cell line: A cell line arises from a primary culture at the time of the first successful subcultures. The term cell line implies that cultures from it consist of numerous lineages of cells originally present in the primary cultures. The terms finite or continuous are used as prefixes if the status of the culture is known. If not, the term line will suffice. The term "continuous line" replaces the term "established line". In any published description of a culture, one must make every attempt to publish the characterization of the history of the culture. If such has already been published, a reference to the original publication must be made. In obtaining a culture from another laboratory, the proper designation of the culture, as originally named and described, must be reported in any publication.

Cell strain: A cell strain is derived either from a primary culture or a cell line by the selection or cloning of cells having specific properties or markers. These properties or markers must persist during subsequent cultivation. In describing a cell strain, its specific features must be defined. The terms finite or continuous are used as prefixes if the status of the culture is known. If not, the term strain will suffice. The term "continuous line" replaces the term "established line". In any published description of a cell strain, one must make every attempt to publish the characterization of the history of the strain. If such has already been published, a reference to the original publication must be made. In obtaining a culture from another laboratory, the proper designation of the culture, as originally named and described must be maintained and any deviations in cultivation from the original should be reported in any publication.

Centrifugal: Developing first at the centre and then gradually toward the outside.

Centripetal: Developing first at the outside and the gradually towards the centre.

Cernuous: Nodding.

Certified seed: Progeny of the registered seed that is produced in the largest volume and sold to the crop producers.

Chaff: Dry, membranous scales or bracts, that is, similar to the chaff which comes from a thrashing machine.

Chaffy: Resembling chaff.

Chamber: A room. Applied to the cavities of an anther or an ovary.

Channeled: With lengthwise grooves, the grooves usually deep.

Chartaceous: Like writing paper.

Chasmogamy: Pollination only after the opening of the flower.

Chimeras: When a mutation occurs within a single cell of a clone it initially produces an "islands" of mutant cells within a growing point of a stem. If successful, the plant then becomes a mixture of two different genotypes. This structural arrangement is known as a chimera.

Chlorophyll: The green colouring of most plants. A substance which aids in making the energy of light available to photosynthesis.

Chloroplast: Literally a green body, actually a solid body with chlorophyll in its outer part.

Chlorosis: A diseased condition shown by loss of green colour.

Choripetalous: With the petals separate or atleast with some of them separate from others.

Cilia: Hairs along the margin of a structure, placed like the eye-lashes on a human eyelid.

Ciliate: With cilia along the margin.

Ciliolate: Diminutive of ciliate.

Cinereous: The colour of ashes.

Circinate: Coiled at the tip, as the young coiled leaf of a fern which uncoils gradually as it develops.

Circumscissile: Opening by a horizontal circular line, the top coming off like a lid.

Class: The next taxon below division; a group of related orders or sometimes a single order.

Clavate: Gradually enlarged upward after the manner of a base-ball bat or the traditional giants club, that is, either tapering gradually upward or with an enlarged knob at the summit.

Clavellate: Diminutive of clavate.

Claw: The narrow stalk at the base of a petal. This resembling the petiole of a leaf.

Cleft: Indented about half way or little more than half way to the base or to the midrib.

Cleistogamy: Self-pollination without the flower opening.

Cleistogamous: Self-fertilized in the bud stage or atleast without opening.

Climbing: Supported by clinging (Scandent).

Clonal propagation: A sexual reproduction of plants that are considered to be genetically uniform and originated from a single individual or explant.

Clone: Genetically uniform material derived from a single individual and propagated exclusively by vegetative means as cuttings, graftage or division.

Coalescence: The union of similar parts as for example, the petals of a flower.

Coalescent: Adjective.

Cochleate: Spiral, like a turban snail.

Coherent: Grown fast together.

Colonial: In colonies. For the angiosperms used, primarily in reference to plants occurring in clumps connected by rhizomes.

Coloure: The failure of blossoms to set, resulting in a pre-matured drop.

Column: A slender aggregation of coalescent stamen. Filaments, as in some members of the mallow family.

Coma: A tuft of hairs on the end of a seed.

Commissure: The surface along which two or more locules are joined to each other.

Comose: With a coma.

Compatible: Ability of pollen to develop in the styles rapidly enough to reach the ovule in time to effect fertilization.

Compatibility: Of sex cells, the ability to unite and form a fertilized egg that can grow to maturity.

Complete flower: A flower with all the four usual series that is sepals, petals stamen and pistils.

Compound: Composed of two or more similar elements.

Compound leaf: It is composed of two or more leaflets, a compound pistil is composed of two or more coalescent carpels.

Compressed: Flattened, particularly from side to side.

Conduplicate: Folded together lengthwise.

Cone: A reproductive structure composed of an axis bearing sporophylls or other seed or pollen bearing structures.

Congeniality: As determined by the degree of success of the union between stock and scion.

Conglomerate: Densely aggregated; a cluster or a heap.

Coniferous: Cone bearing.

Conjugate: Lying together in pairs.

Connate: Joined from the beginning of development.

Connate perfoliate: Both connate and perfoliate.

Connective: The middle part of an anther connecting the two pollen sacs or two pairs of pollen sacs.

Connivent: Standing together. Stamens with their tips ending against each other are connivent.

Contorted: Twisted out of the usual form. Petals may be contorted in the bud.

Convolute: Rolled up lengthwise. Petals may be convolute in the bud.

Cordate: Of a conventional heart-shape, the length greater than the width, the petiole attached in the basal sinus; applied also to the basal indentation of a leaf or other structures.

Coriaceous: Leathery.

Corm: A bulblike structure formed by enargement of the stem base. It is sometimes coated with one or more membranous layers.

Corneous: Horny.

Corolla: The petals of a flower, that is, the inner series (or several series) of the parianth. Each petal has usually a single vascular trace as does a stamen.

Corolla division: A division is longer than a lobe or part.

Corolla lobe: A lobe is longer then the tooth. Lobes may be applied in a broad sense to cover lobes, parts or divisions.

Corolla part: A part is longer than a lobe.

Corolla teeth: The separate tips of the individual petals of a sympetalous corolla.

Corolla tube: The hollow cylinder formed by coalescence of petals.

Corona: A small crown, the term being applied to special appendages of the corolla. These sometimes form a tube.

Coroniform: Shaped like a crown.

Corrugated: With many small folds or wrinkles.

Cortex: The layers of living cells outside the central stele but inside the outermost single layer of cells of a stem or a root.

Cortical: Of the cortex.

Corymb: A flat topped cluster of flowers; fundamentally like a raceme but with the pedicels of the lower flowers longer and the pedicels of the upper flowers gradually shorter.

Corymbiform: In the form of a corymb.

Corymbose: Arranged in corymbs.

Costa: A rib or a promiment nerve.

Costate: Ribbed lengthwise.

Cotyledons: One of the first leaves developed in the embryo in the seed at the joining point of the hypocotyls and the epicotyl. Often food materials are stored in them.

Creeping: The stem growing along the ground and producing adventitious roots.

Crenate: With rounded teeth projecting at right angles to the edge of the leaves.

Crenulate: Diminutive of crenate.

Crested: With a crest, that is a projection at a prominent position such as the apex or the midrib of an organ.

Crinkle: A disorder of apples in which the surface of the fruit becomes roughened. Supposed to be a form of draught injury.

Crisped: Ruffled, that is with the same bind of winding as in sinuate, but in the vertical plane instead of the horizontal.

Cristate: Crested.

Cristulate: The diminutive of cristate.

Cross-pollination: Transfer of pollen grain from a flower of one variety to a flower of another variety.

Cross-compatible: The pollen of one variety 'A' is capable of functioning in the styles and fertilizing the ovules of variety 'B' (variety 'B', however, may not be cross-compatible with 'A').

Cross-incompatible: Variety 'A' produces functional sex cells, but its pollen tube grows too slowly in the styles of variety 'B' to effect fertilization. Variety 'A' may, however, serve as an effective pollinizer for some other varieties. (Variety 'B' mar serve as an effective pollinizer for variety 'A').

Cross-fertile: One variety 'A' is used as a pollinizer for another variety 'B' and 'B' produces fruit with viable seed.

Cross-fruitful: One variety 'A' is used as a pollinizer for another variety 'B' and 'B' produces a commercial crop.

Cross-pollination: The transfer of pollen from the anthers of a flower of one variety to the stigma of a flower of a different variety.

Cross-unfruitful: One variety 'A' is used as a pollinizer for another variety 'B', and 'B' fails to produce a commercial crop.

Cross-sterile: One variety 'A' is used as a pollinizer for another variety 'B', and 'B' fails to produce fruit with viable seed.

Cross section: A slice cut across an object.

Cruciate, cruciform: Cross like.

Crustaceous: Hard and brittle in texture.

Cryptogames: An old term applied to the thallophytes, bryophytes and the pteridophytes, that is the plants which regularly reproduce with out formation of seeds.

Cucullate: Hood shaped

Culum: The hallow stem of a grass, the term being applied some times to ridges also.

Cuneate, cuneiform: Wedge shaped; essentially a narrow isosceles triangle with the distal corners rounded off. The petiole of a cuneate leaf is attached at the sharp angle.

Cupule: A little cup.

Cuspidate: With a sharp, firm point at the tip, as applied to leaves, the term indicates a point of firmer texture than the rest of the blade.

Cuticle: The waxy, more or less waterproof coating secreted by the cells of the epidermis of a leaf, stem or flower parts.

Cyathiform: Cup shaped.

Cyathium: A cuplike involucre enclosing flowers.

Cycle: A circle. Leaves or flower parts arranged in a single series at a single node or in a cycle.

Cyclic: Arranged in a cycle.

Cylindroidal: In approximately the shape of a cylinder.

Cyme: A broad, more or less flat-topped cluster of flowers.

Cymose: With cyme like cluster of flowers.

Cymule: Diminutive of cyme.

Cypsela: An achene which is attached to the enclosing floral cup or tube, that is, an achene formed from an inferior ovary.

Cytology: The study of cell. Often particular emphasis is placed upon the chromosomes.

Cytoplasm: The living substance outside the nucleus and inside the enclosing membrane.

Deciduous: Falling off at the end of each growing season.

Deciduous tree: Tree which shed their leaves during winter months to enable them to stand cold injury.

Declinate: Bending or curving downward or forward.

Declined: Bending over in one direction.

Decompound: More than one compound.

Decombent: Reclining except at the apex.

Decurrent: Leaf bases which continue along the stem as wings or lines.

Decussate: In opposite pairs with the alternate pairs projecting at right angles to each other.

Deflexed: Abruptly bent or turned downward.

Defined medium: A nutritive solution for culturing cells in which each component is of known chemical structure. Although it is recognized that even the "Purest" chemical compounds may have some contaminants, high quality chemicals should be used with analytic data, if possible, on contaminants.

Dehisce: The split open along definite lines.

Dehiscence: The process of splitting open at maturity.

Dehiscent: Splitting open along definite lines.

Deltoid: Of the shape of the greek letter delta, that is, an equilateral triangle, the attachment being in the middle of one side.

Dentate: With angular teeth projecting at right angles to the edge of the structure.

Denticulate: Diminutive of dentate.

Depauperate: Stunted, that is, very small.

Depressed: Flattened from above as if pushed downward.

Determinate: With a definite predetermined number of structures.

Diadelphous: In two brotherhoods, the term being applied to stamens coalescent in two sets.

Diandrous: With two stamens.

Dichasium: A cyme with two axis running in opposite directions, that is, the type of cyme formed in plants with opposite branching in the inflorescence. See cyme.

Dichlamydeous: With both calyx and corolla.

Dichogamy: Insuring cross-pollination by the sexes being developed at different times.

Dichotomous: Forking, with two usually equal branches at each point of forking.

Dicotyledonous: With two cotyledons.

Dicliny: Male and female organ separate and in different flowers.

Didymous: Twin like, that is, occurring in pairs.

Didynamous: In two pairs, the pairs not being of the same length. The term is applied to stamen.

Diffuse: Spreading widely and diffusely in all directions.

Digitate: Resembling the fingers of a human hand, that is, with several similar structures arising at a common point. See palmate.

Digynous: With two pistils.

Dimerous: With two members; the flower parts in two.

Dimidiate: With the appearance of being only half a structure.

Dimorphous: With two forms.

Dioecious: Unisexual, the male and female elements in different individuals.

Diploid: With 2n chromosomes per cell.

Diplostemonous: The stamens in two series, those of outer series alternating with the petals.

Dipterous: With two wings.

Disc, disk: In the compositae, the central portion of the compound receptacle bearing the disc flowers.

Disc flowers: One of the flowers with tubular corollas.

Disciform: Circular and flattened like disc.

Discoid: Resembling a disc.

Discrete: Separate.

Dissected: Divided into narrow segments.

Dissepiment: A portion dividing an ovary or a fruit into chambers.

Distichous: In two vertical series.

Distinct: Separate.

Diurnal: Occuring in the day time. The term is applied to the flowers which open in day light as opposed to the nocturnal flowers which open at night.

Divaricate: Spreading widely, that is, divergent.

Divergent: Spreading away from each other.

Divided: Indented essentially to the base or midrib.

Division: The highest rank of taxon in the plant kingdom, a group of related classes or some times a single class; a segment of a structure, such as leaf.

Dormant: Applied to buds when they are not actively growing and to plants when they are not in leaf.

Dorsal: On the outer surface of an organ, that is, the side away from the axis.

Dorsiventral: A structure having a clear differentiation of a back and front or upper and lower side.

Dorsoventral: The distance or measurement from the back of a structure to the front, that is, from dorsal to ventral, as opposed to lateral distance.

Double samara: A Samara with two locules and two wings.

Downy: Finely and softly pubescent.

Drainage area: The watershed or area into which the excess precipitation of a region drains.

Drupaceous: Druplike.

Drupe: A fruit with a fleshy exocarp and a hard, stony endocarp about each seed. The term is used also for other fruits which are fleshy or pulpy on the outside and with one or more stones inside.

Drupelet: Diminutive of drupe. The cluster of fruits of a blackberry or raspberry is composed of druplets which seem to form a single fruit.

Echinate: covered with prickles.

Egg: A female gamete cell of a plant.

Effective bloom: The length of time the tree is in conspicuous blossom.

Ellipsoid, ellipsoidal: Elliptical in outline with three-dimensional body.

Elliptic, elliptical: In the form of an ellipse, that is, about one and one half times as long as broad, widest at the middle and rounded at both ends. A two dimensional figure.

Emarginate: With a shallow broad notch at the apex.

Embryo: The new plant enclosed in the seed. Its formation follows union of gametes.

Embryo culture: In vitro development or maintenance of isolated mature or immature embryos.

Embryogenesis: The process of embryo initiation and development.

Embryo sac: The cell in the ovule in which the embryo is formed.

Emersed: Above water.

Enation: An outgrowth from the superficial tisssues of the stem.

Endimic: Restricted in occurrence to aparticular geographical area.

Endocarp: The inner layer of the wall of a fruitified ovary.

Endosperm: A cell layer occurring in at least the immature seeds of flowering plants.

Ensiform: Sword like; in the form of sword.

Entire: without division or toothing of any bind.

Entomophilous: Insect pollinated.

Emphemeral: Lasting for a brief period.

Epicotyl: The portion of the embryo of a seed plant just above the cotyledons, the young stem.

Epigynous: The sepals, petals and stamens apparently upon the ovary, but actually growing from the edge of the floral cup, which is adnate to the ovary.

Epigynous disc: A disc within the floral cup and on top of the inferior ovary.

Epiphyte: A plant growing upon another plant but not parasitic upon it.

Equitant: With the leaves folded around a stem after the manner of the legs of a rider around a horse or some times with the leaves folded around each other in rows.

Errect: Standing upright.

Erose: With the margin appearing to have been gnawes.

Etiolated: White through failure to develop chlorophyll.

Euploid: The situation which exists when the nucleus of a cell contains exact multiples of the haploid number of chromosomes.

Even pinnate: Without a terminal leaflet.

Exocarp: The outer layer of the wall of a fruitified ovary.

Excentric: Off centre.

Excortis: A shelling off the bark of tree.

Excurrent: Projecting as for example, a leaf base which projects beyond the margin of the blade.

Exfoliating: Separating into thin layer.

Exocarp: The outer layer of the pericarp.

Explant: Tissue taken from its original site and transferred to an artifical medium for growth and maintenance.

Explanate: Flattened and spread out.

Exserted: Projecting beyond the usual containing structure.

Extraterritorial: Occuring beyond the specific geographical range of some features of this book ie. beyond the north America north of Mexico.

Extrose: Facing outward.

Falcate: In the shape of a scythe, that is curving, flat and tapering gradually to a point.

False partition: An ovary partition not formed by infolding of the edges of the carpel, growing from the middle of the placenta on the opposite side or formed by ingrowth of placentae.

Farinose: Mealy, that is composed of mealy granules or with granules on the surface.

Fasciate: Literally, in a bundle or bundled together, the term being applied as fascinated, to branches remaining parallel and grown abnormally together.

Fasciation: A deformity in which the stem becomes much flattened as a result of multiple terminal buds arranged in a single plane.

Fascicle: A bundle or cluster.

Fastigiate: Errect and close together.

Faveolate or favose: Resembling a honey comb.

Fecundation: The fusions of two gametes to form a new cell.

Fecundity: The ability of flowers to produce seeds that will germinate.

Female gamete: An egg; a female gamete cell.

Fenestrate: Perforated.

Ferruginous: Rust coloured.

Fertile: Productive, mean capable of producing fruit or spores. Sometimes a staminate flower is referred to as infertile or sterile because it produces no seed.

Fertility: The ability to set and mature fruit with viable seed.

Fertilization: The union of male germ cell, contained in the pollen grain, with the female germ cell, or egg, in the ovule.

Fibrillose: With fine fibres.

Fibrous root system: A root system with several major roots about equal and arising from approximately the same point.

Field capacity: Is defined as the amount of water held against the force of gravity.

Filament: A thread; the stalk of a stamen.

Filamentous: Composed of threads.

Filiform: Thread like, that is, long slender and cylindroidal.

Fimbriate: Fringed, that is resembling the fring on the sleeve of an early American buckskin shirt.

Fimbrillate: The diminutive of fimbriate.

Fimbriolate: With a very fine fringe.

Fistulose: Hollow and cylindroidal.

Flabellate: Fan shaped.

Flaccid: Weak.

Flagelliform: Whiplike or lash like.

Flates: They are shallow plastic, styrofoam, wooden or metal trays with drainage holes in the bottom. They are useful for germinating seeds or rooting cuttings, since they permit young plants to be moved easily.

Fleshy fruits: A fruit with soft, juicy tissues.

Flexuous: Curved in first one direction, then the opposite.

Floccose: With tufts of wooly hair.

Flocculent: Diminutive of floccose.

Floral cup: A cup bearing on its rim the sepals, petals and stamens; originating as (a) a hypanthium. (b) a "calyx tube" formed by coalescence and adnation of the bases of the sepals, petals and stamens, or (c) a perianth tube of the type in the monocotyledons, the sepals and adjacent petals being adnate edge to edge.

Floral tube: An elongated, slender floral cup.

Floret: A small flower, the flower of a grass and the two immediately enclosing bracts, that is the lemma and palea.

Floricana: A canelike stem producing flowers.

Floriferous: Bearing flowers.

Flower: A complex strobilus formed at the end of a branch, including the receptacle (thalamus or torus) and bearing sepals, stamens, petals and pistils or some of these.

Flower bud: A bud which will develop into a flower, actually the young flower.

Flowering hormones: Are hormones which initiate the formation of floral primodia, or promote their development.

Flowering regulators: Are regulators which affect flowering.

Foliaceous: Leaflike.

Foliage bud: A bud containing new young leaves.

Foliage leaf: The green leaf of the mature plant.

Foliar: Relating to a leaf.

Foliate: Leaved, an example being 3- foliate or trifoliate.

Foliolate: With leaflets.

Foliose: Bearing many leaves.

Follicle: A dry fruit formed from a single carpel, containing more than one seed, and splitting open along the suture. The term is applied also sometimes to similar fruits splitting only along the midrib.

Follicular: Of the nature of a follicle.

Forked: Dividing into branches which are nearly equal.

Foundation seeds: Progeny of breeders seed which is so handled as to maintain the highest standard of genetic identity and purity.

Foveolate: Pitted.

Free: Separate from other organs (distinct).

Friability: A term indicating the tendency for plant cells to separate from one another.

Frond: The leaf blade of a fern.

Frondose: Leafy, that is with frond like leaves.

Fruit: A matured ovary with its enclosed seeds and some times with attached external structures.

Fruitescent: Shrubby or becoming so at length.

Fruitful: A plant or variety that sets and matures a commercial crop of fruit.

Fruiticose: Distinctly woody, living over from year to year, and attaining considerable size, with several main stems instead of a single trunk.

Fruit setting: A development of the ovary and adjacent tissues following the blossoming period.

Fugacious: Falling away early.

Fulbous: Tawny.

Fungus: A plant without chlorophyll and with no vascular tissue, stems or leaves. Plural fungi.

Funiculus: The stalk of an ovule or a seed.

Funnelform: In the shape of a funnel.

Furcate: Forked.

Fusicous: Grayish-brown.

Fusiform: In the shape of a spindle, that is, widest at the middle and tapering gradually to each pointed end, the body being circular in cross section.

Galea: A hood formed from a portion of the parianth derived, for example, from the two upper petals coalascent indistinguishably into one or from the upper sepal.

Galeate: With a galea.

Gamete: A unisexual cell which must fuse with another gamete to produce a new individual.

Gamopetalous: With all the petals coalescent (Sympetalus).

Gametophyte: The gamete producing generation, each cell with a chromosomes.

Geminiate: Equal or arranged in pairs.

Generative cell: A cell in the microgametophyte (or pollen grain) which may give rise directly or indirectly to male gamete cells or nuclei.

Genetics: The study of heredity.

Geniculate: Bent like the human knee.

Genus: A group of related species or sometimes a single species. Plural, genera.

Gibbous: Swollen or distended on one side.

Glabrate: At first hairy but latter becoming glabrous.

Glabrous: not hairy.

Gladiate: Sword like.

Gland: A secreting organ.

Glandular: Bearing glands, small cellular organs secreting oils, tars, resins etc. often only the secretion is visible.

Glaucescent: More or less glaucous.

Glaucous: Covered with a white or bluish powder or bloom, this often composed of finely divided particles of wax. Ex. The bloom on a plum.

Globose, globular: Spheroidal.

Glochid: A sharp hair or bristle tipped with a barb.

Glochidiate: Barbed at the tip.

Glomerate: In compact clusters.

Glomerulate: Diminutive of glumerate.

Glomerule: A single cyme of sessile flowers forming a compact cluster. It may be distinguished from a head by blooming of the central flower first.

Glumaceous: Resembling a glume of a grass spikelet.

Glume: One of the two chaff like bractlets at the base of a grass spikelet. The glumes do not enclose flowers.

Glutinous: covered with sticky material.

Granulose: Covered by minute grains of hardened material.

Gregarious: In large colonies.

Growth hormones: A hormone which regulate growth.

Growth regulators: (Synonym: growth sunstances) are regulators which affect growth.

Gummosis: A disorder particularly of stone and citrus fruit, in which there are copious exudations or deposites of gum.

Gum spots: A disorder of stone, citrus and certain other fruits in which there are small local deposites of gum in the tissues of fruits, shoot or other organ.

Gymnospermous: With the seeds naked, that is not enclosed in an ovary.

Gynandrium: A structure formed by adnation of stamens and pistil.

Gynandrous: With the stamen adhering to the pistils.

Gynecandrous: With staminate and pistillate flowers in the same inflorescence, the pistillate above and the staminate below.

Gynobase: An enlarged or elongated portion of the receptacle bearing the pistil.

Gynoecium: The pistil of a flower or its group of pistils.

Gynophore: A special stalk under a pistil.

Habit: The general appearance of a plant.

Habitate: The type of locality or the set of ecological conditions under which the plant grows.

Hair: A slender cellular projection.

Halberd-shaped: Hastate.

Halophyte: A plant growing in salty soil as for example an alkali flat or a salt water marsh.

Hamate: With a hook at the tip.

Haploid: With a chromosomes per cell.

Hastate: More or less sagittate (arrow head shaped) but with the divergent basal lobes.

Head: A cluster of sessile or essentially sessile flowers or fruits at the apex of a peduncle. Essentially a spike with a very short axis.

Helicoid: Spiral, like the shell of a turban snail.

Helicoid cyme: A coiled inflorescence. The main stem terminates in a flower and a single bud just below grows out as a stem but terminates also in a flower, the process being repeated many times and always in the same direction.

Herb: A non woody plant, at laeast one which is not woody above ground level.

Herbaceous: Not woody.

Herbarium: A collection of pressed plant specimens.

Hermaphrodite: Bisexual, that is, with stamens and pistils in the same flower.

Heterocarpous: Producing more than one kind of fruit.

Heterogamous: With more than one kind of flower.

Heterogeneous: Of various binds, that is, not all individuals the same.

Heteroploid: The term given to a cell culture when the cells comprising the culture possess nuclei containing chromosome numbers other than the diploid number. This ia a term used only to describe a culture and is not used to describe the individual cells. Thus, a heteroploid culture would be one which contains aneuploid cells.

Hilium: The point of attachment of the funiculus (Stalk) to the seed.

Hip (of a rose): A floral cup which usually becomes enlarged and fleshy at fruiting time. Thetrue fruits are achenes inside.

Hippocrepiform: Horse shoe shaped.

Hirsute: With fairly coarse more or less stiff hairs.

Hirsutulus: Slight hirsute.

Hispid: With rigid or stiff bristles or bristly hairs.

Hispidulous: Diminutive of hispid.

Hoary: Covered with short, dense, grayish white hairs, the surface of the stem, leaf or other structure therefore appearing white.

Homogamous: With only one kind of flower.

Homogeneous: With all members similar.

Hood: A hoodlike structure, often formed from a petal or a sepal or more than one of either.

Humifuse: Spreading on the ground.

Hyaline: Thin and membranous being transparent or translucent.

Hybrid: Produced by dissimilar parents.

Hydrophyte: An aquatic plant.

Hygroscopic: Changing form with changes of moisture content.

Hygroscopic coefficient: The percentage of soil water retained in contact with a saturated atmosphere and in the absence of any other source of moisture.

Hypanthium: A floral cup or tube developed by extra growth of the margin of the receptacle. An alternative adopted by some authors in use of hypanthium with the meaning of floral cup.

Hypocotyl: The portion of the axis of an embryo of a seed plant just below the cotyledon(s).

Hypogaeous: Below the ovary, that is, referring to flower parts which come directly from the receptacle and not from a floral cup or tube. A hypogynous flower is one which does not have a floral cup or tube.

Hypogynous disc: A flushy cushion of tissue growing from the receptacle below the ovary, often the stamens and sometimes the petals being attached to it.

Imbricate: overlapping like the shingles on a roof.

Imparipinnate: Odd pinnate.

Implexed: Tangled, interlacing.

Implicate: Woven in.

Inactive bud: A bud dormant either before the beginning of the growing season or through a longer period.

Incanous: Covered with white hairs.

Included: Not protruding beyond the normal surrounding structures.

Incompatible: Inability of viable pollen to develop in the styles rapidly enough to reach the ovule in time to effect fertilization.

Incompatibility: Of sex cells, the inability to unite and form a fertilized egg that can be grown to maturity.

Incomplete flower: A flower lacking one or more of the four usual series of structures, that is, sepals, petals, stamens or pistils.

Incrassate: Thickened.

Incumbent: Applied to a pair of cotyledons which lie with the back of one against the axis of the embryo.

Indehiscent: Not opening by splitting along regular lines or not opening at all.

Indeterminate: Of indefinite growth ie. the size or number of organs to be produced not predetermined.

Indigenous: Native in a particular region.

Indument: A covering of hairs.

Induplicate: With the edges folded inward.

Indurate: Hardened.

Indusium: The membranous covering of a sorus of a pteridophyte.

Inferior: Below.

Inferior floral cup: The portion of a floral cup or tube adnate to the wall of an inferior ovary.

Infertile: Varieties 'A' and 'B' both produce fruit with viable when pollinated by each other.

Inflorescence: The flowering area or segment of a plant.

Integument: The outer coating of an ovule, a structure grown up from the sporophyll and surrounding the sporangium. The integument becomes the seedcoat.

Intercompatible: The pollen produced by either variety of a combination is capable of functioning in the styles and fertilizing the ovule of the other variety.

Interfruitful: Varieties 'A' and 'B' both produce commercial crops when pollinated by each other.

Interincompatible: Varieties 'A' and 'B' are unfruitful when pollinated by each other because the pollen tubes of each variety grow too slowly in the styles of the other to effect fertilization. Either variety may serve as an effective pollinizer for some other varieties.

Intersterile: Varieties 'A' and 'B' both fail to produce fruit with viable seed when pollinated by each other.

Interstock: It is a piece of stem inserted by means of two graft unions between the scion and the rootstock.

In vitro propagation: Propagation of plants in controlled, artificial environment, using plastic or glass cultures vessels, aseptic techniques, and a defined growing medium.

In vitro transformation: A heritable change, occurring in cells in culture, either intrinsically or from treatment with chemical carcinogens, oncogenic viruses, irradiation, etc., and leading to the acquisition of altered morphological, antigenic, neoplastic, proliferative, or other properties.

Involucel: The involucre of a secondary umbel.

Involucre: A series of bracts surrounding a flower cluster or some times a single flower.

Irregular: Used in botany to indicate a bilaterally symmetrical structure as for example, a bilaterally symmetrical flower.

Isomerous: With the members of the various series of flower parts of equal number.

Keel: A ridge along the outside of a fold, like the keel of a boat. The term is applied to the two coalescent lower petals of a papilionaceous corolla of the pea family.

Labiate: With lips, that is, two opposed structures as for example, the upper two petals as opposed to lower three petals in the mint family.

Lanolin: The natural fat of wool.

Latent bud: A bud, usually concealed, more than one year old, which may remain dormant indefinitely or may develop under certain conditions.

Lateral bud: One in a leaf axil.

Legume: A dry several seeded fruit formed from a single carpel and dehiscent on both margins.

Locular: Divided into locules.

Loculicidal: Dehiscent along the midrib of a carpel of an ovary containing more than one carpel and chamber.

Lodicule: One of the two small scales at the base of the grass flower.

Loment: A legume divided by constriction into a linear series of segments each containing one seed.

Male gamete: An antherozoid or a male gamete cell or nucleus.

Male nucleous: The male gamete of a flowering plant, that is, a single nucleus with the associated cytoplasm but with no cell wall.

Megagametophyte: A female gametophyte.

Megasporangium: A sporangium producing megaspore.

Megaspore: A larger type of spore developed by a plant with spores of two sizes or kinds.

Megasporophyll: A sporophyll bearing one or more megasporangia.

Mericarp: A portion of a fruit which appears to be a whole fruit.

Mericloning: A popular term, not in scientific usage, referring to the in vitro vegetative propagation of orchids from excised shoot tips, axillary buds, or floral organs.

Meristem: A plant tissue in which the cells are capable of, under favourable growth conditions usually in process of active division, to produce new daughter cells.

Meristem culture: In vitro culture of a generally shiny dome-like structure measuring less than 0.1 mm in length when excised, most often excised from the shoot apex.

Meristemming: A popular term, not in scientific usage, referring to the in vitro clonal propagation of plants from various explant sources including shoot tips, leaf section and calli; shoot apical meristems are seldom used.

Meristemoid: Meristem-like cells located in areas of a plant or culture other than the meristem.

Mesocarp: The middle layer of a pericarp.

Metaxenia: The supposed direct influence of the pollen or pollen parent on the characteristics of the developing fruit.

Micropropagation: This term is synonymous with in vitro propagation.

Morphogenesis: The process of growth and development of differentiated structures.

Microspore: The smaller type of spore in a plant producing spores of two sizes.

Mixed bud: A bud containing both leaf primordial and rudimentary flowers.

Monoadelphous: In one botherhood; referring to stamens with their filaments coalescent into a single tube.

Monochlamydeous: With a single series in the perianth that is only the sepals.

Monoecious: The stamens and pistils in separate flowers borne on the same individual.

Naked bud: A bud not covered by special scales but only by the outer leaves.

Nocturnal: Occuring at night. The term is applied to flowers which open at night.

Nucellus: The sporangium enclosed in the ovule of a seed plant.

Nut: A hard, relatively large, indehiscent, 1 seeded fruit.

Nutlet: Diminutive of nut.

Nutrients: Are defined as materials which supply either energy or essential mineral elements.

Order: A tascon composed of related families or sometimes of a single family.

Organ culture: The maintenance or growth of organ primordial or the whole or parts of an organ in vitro in a way that may allow differentiation and preservation of the architecture and/or function.

Organogenesis: The evolution, from dissociated cells, of a structure which shows natural organ form or function or both.

Osmosis: The passage of water through a semi-permeable membrane.

Ovary: In the flower, that portion of the female organ which contains the potential seeds (ovules). After fertilization it grows into the fruit.

Ovulate: Producing ovules.

Ovules: The potential seed before fertilization.

Panicle: A cluster of associated spikes racemes or corymbs.

Papilionaceous: Butterfly like.

Pappus: The specialized calyx of members of the sunflower family; composed of bristles or scales.

Parthenocarpy: The development of edible fruit without fertilization. Parthenocarpic fruits are seedless.

Parthenogenetic: Developing without fertilization

Partly inferior ovary: With only the basal portion of the ovary adnate to the floral cup.

Passage: The transfer or transplantation of cells, with or without dilution, from one culture vessel to another. It is understood that any time cells are transferred from one vessel to another, a certain portion of he cells may be lost and, therefore, dilution of cells, whether deliberate or not, may occur. This term is synonymous with the term "subculture".

Passage number: The number of times the cells in the culture have been subcultured or passaged. In descriptions of this process, the ratio or dilution of the cells should be stated so that the relative cultural "age" can be ascertained.

Peat: It consists of remains of a aquatic, marsh, bog or swamp vegetation which has been preserved under water in a partially decompose state.

Pectinate: Resembling a comb.

Pedicel: The internode below a flower.

Peduncle: The stalk of a cluster of flowers or the next to the last internode below a single flower.

Pentagynous: With five pistils.

Pentandrous: With five stamens.

Pepo: A fruit of the type in the gourd family, that is, with a hard or leathery rind on the outside and fleshy placental tissue inside. The seeds are numerous and there is only one chamber.

Perfect: A flower having both functional pistils and functional stamens.

Perennation: A lasting state referring particularly to the persistence of fruit long after its usual season of maturity.

Perianth: A collective term for the calyx and the corolla.

Perianth tube: A tube formed by coalescence and adnation of the lower portions of the sepals and petals.

Pericarp: The wall of a matured ovary, that is, the wall of the fruit or the inner wall if the ovary is inferior.

Perigynium: A special sac which encloses the ovary of calyx.

Perigynous: Descriptive of the parts of a flower attached to a floral cup or tube which is not adnate to the ovary. A perigynous flower has a floral cup but the ovary is not adnate to it.

Perigynous disc: A disc adnate to the floral cup of a perigynous flower.

Perlite: It is a graywhite silicaceous material of volcanic origin, mined from larva flows.

Permanent wilting point: The amount of water in soils when rapidly growing plants fail to recover from wilting under condition of low transpiration.

Petal: One member of the series of flower parts forming the corolla.

Petaloid: Resembling a petal in colour and texture.

Photosynthesis: The process of combining in the presence of chlorophyll and light of carbon dioxide and water yielding glucose and oxygen.

Photosynthetic: Descriptive of a green portion of a plant capable of carrying on photosynthesis.

Pistil: The female organ of a flower.

Pistillate: Having pistils, that is, a flower which has pistils but no stamens.

Placenta: The tissue of an ovary which bears ovules or seeds.

(Plant) Hormones: (Synonym: phytohormones) are regulators produced by plants, which in low concentrations regulate plant physiological processes. Hormones usually move within the plants from a site of production to a site of action.

(Plant) Regulators: Are organic compounds, other than nutrients which in small amount promote, inhibit or otherwise modify any physiological process in plants.

Plating efficiency: The percentage of cells plated which give rise to colonies. The total number of cells in the inoculum, type of culture vessel and the environment conditions (medium, temperature, closed or open system, etc.) must always be stated. This term is often expressed as the percentage of individual cells in the vessel which give rise to colonies. If one is certain that each of the colonies arose from single cells, then one may properly apply the term "cloning efficiency".

Pollen: The spheroidal structures produced in an anther of a flower or in the microsporophyll of a gymnosperm. The pollen grains are microgametophytes developed from microspores.

Pollen sac: Pollen bearing cavity.

Pollen tube: A tube developed by a pollengrain and growing through the microphyle of an ovule where its contained cells or nuclei effect fertilization of the egg.

Pollination: The transfer of pollen to the stigma; or in a broad sense, the distribution of pollen. Pollination may be accomplished by insects, wind, gravity, water, birds and artificial methods derived by man.

Pollinizer: The variety (plant) used to furnish pollen. The male parent.

Polyadelphous: In several brotherhoods; with several groups of coalescent stamens.

Polyandrous: With an indefinite large number of stamens.

Polyembryony: Production of more than one embryo in a single ovule.

Polygamo-dioecious: With hermaphrodite and unisexual flowers on different individuals of the same species.

Polygamo-monoecious: Polygamous but in the main monoecious.

Polygamous: With bisexual and unisexual flowers on the same or different individuals of the species.

Polygynous: With a large indefinite number of pistils.

Polymorphic: With several or many forms.

Polyploid: With several to many times n chromosomes per cell.

Population density: The number of cells per unit area or volume of a culture vessel. Also, the number of cells per unit volume of medium in a suspension culture.

Pome: A fleshy fruit with several seed chambers, this formed from an inferior ovary, the fleshy tissue being largely the floral cup, the seeds not embedded in pulp. Apples and pears are the classical examples.

Prickle: A sharp, pointed outgrowth from the superficial tissues of the stem, that is, from the epidermis or the cortex, as in a rose prickle. The structure is not associated with the conducting tissues at the centre of the stem.

Primary culture: A culture started from cells, tissues or organs taken directly from organisms. A primary culture may be regarded as such until it is successfully subcultured for the first time. It then becomes "cell line".

Protandry: The pollen being discharged before the pistils are receptive.

Protogyny: The pistils being receptive before the anthers have ripe pollen.

Pruinose: Covered with a bloom, that is, finely divided particles of a waxy powder.

Protoplast: A plant cell without its outer retaining cell wall.

Protoplast fusion: Technique in which protoplasts are fused into a single cell.

Pruning: Means removing certain parts of the tree in order to modify and utilize its natural habits so that more and better fruit can be obtained at less cost over a longer period.

Pumice: Increases aeration and drainage in a propagation mix and can be used alone or mixed with peat moss.

Raceme: An inflorescence composed of pedicelled flowers arranged along an axis which elongates for an indefinite period. The lower flower blooms first and eventually the terminal bud forms the last flower.

Racemiform: In the form of a raceme.

Rachis: The axis of a pinnate leaf or of an inflorescence.

Ray: A pedicel within an umbel; a ray flower or its corolla.

Ray flower: In the sunflower family, one of the flowers with a ligulate corolla. The flattening is due to failure of complete coalescence between two petals of the sympetalous corolla.

Receptacle: The apical area beyond the pedicle, that is, the portion which bears flower parts. The receptacle consists of several or many nodes and short internodes.

Regular: Radially symmetrical.

Reproductive bud: A bud developing into a flower, cone or other pollen-, seed- or spore bearing structures.

Resting bud: An inactive bud.

Rockwool: This material is used as a rooting and growing medium.

Rouging: The removal of off-type plants or plants of other varieties.

Rosette: A condition in which the internodes are much shortened giving the leaves a bunched or clustered appearance.

Salverform: Descriptive of a sympetalous corolla with the slender basal tube abruptly expanded into a flat or saucer shaped upper portion.

Samara: A dry indehiscent fruit with a wing.

Sand: It consists of small rock particles 0.05-2.0 mm in diameter, formed as a result of weathering of various rocks, its mineral composition depending upon the type of rock.

Scape: A flowering stem which bears no leaves or only a small bract or a pair or whorl of bracts.

Schizocarp: A fruit which splits into one seeded sections (mericarp).

Scion: It is the short piece of detached, shoot containing several dormant buds which, when united with rootstock, comprises the upper portion of graft.

Scorpioid: Often used as descriptive of an inflorescence which is coiled in the bud stage as in the forget me not or fiddle neck. This is a helicoids cyme although it appears to be a spike or a raceme.

Seed: A mature ovule consisting of an integument (seed coat), an enclosed nucellus, the remains of the megagametophyte, the endosperm and embryo.

Seed coat: The outer coating of a seed developed from the integument of the ovule.

Seed dormancy: It is a condition where seeds will not germinate even when the environmental conditions (water, temperature and aeration) are permissive for germination. This not only prevents immediate germination but also regulates the time, conditions and place that germination will occur.

Self-fertile: The ability of a variety to produce fruit with viable seed following self-pollination.

Self-pollination: The transfer of pollen from the anthers of a flower to one variety to the stigma of a flower of the same variety.

Self-sterile: The inability of a variety to produce fruit with viable seed following self-pollination. (Some varieties may be self-fruitful, even though they are self-sterile, because of their ability to produce parthenocarpic fruits.).

Self-unfruitful: A variety which is unable to set and mature a commercial crop of fruit with its own pollen (or as a result of parthenocarpic fruit development).

Sepal: One of the flower parts of the outer series, the sepal forming a calyx.

Septicidal: Descriptive of a fruit dehiscent through the partitious between the seed chambers.

Silique: The elongate capsular fruit of the mustard family which has two seed chambers separated by a false partition from the middle of one placenta to the middle of other. The false partition in this case results from presence of some fertile and some sterile carpels in the pistil.

Sod culture: A method of orchard soil management in which a permanent perennial crop is grown between the trees. Mowed once or twice during growing season and then allowed to remain on the ground. A limited area around the tree is hoed, spaded or otherwise tilled.

Sod mulch: A method of charged soil management in which a permanent perennial crop is grown between the trees, mowed once or twice during the growing season and then allowed to remain on the ground.

Somaclonal variation: This type of genetic variation occurs when plant cells of certain species were grown in culture and new plant regenerated.

Somatic cell hybrid: The cell or plant resulting from the fusion of animal cells or plant protoplasts, respectively, derived from somatic cells which genetically.

Somatic hybridization: The in vitro fusion of animal cells or plant protoplasts derived from somatic cells which differ genetically.

Spadix: A spike with a fleshy or succulent axis, the flowers often partly embedded in the axis.

Spathe: A large bract enclosing an inflorescence at least when it is young. Spathes are either white or highly coloured but usually not green.

Species: A group of related varieties or often a single unit.

Spike: An inflorescence in which the sessile flowers are arranged along an axis. The basal flower blooms first, the last one formed is at the apex.

Spur: An elongated sac produced from a part of a flower as in the larkspur from a sepal or in the columbines from a petal.

Stage I: A step in vitro propagation characterized by the establishment of an asceptic tissue culture of a plant.

Stage II: A step in vitro propagation characterized by the rapid numerical increase of organs or other structures.

Stage III: A step in vitro propagation characterized by the preparation of propagule for successful transfer to soil, a process involving rooting of shoot cuttings, hardening of plants and initiating the change from the heterotrophic to the autotrophic state.

Stage IV: A step in vitro propagation characterized by the establishment in soil of a tissue culture derived plant, either after undergoing a Stage II pretransplant treatment or, in certain species, after the direct transfer of plants from Stage II into soil.

Stamen: The pollen-producing structure of a flowering plant, consisting of an anther and of a filament.

Staminodium: A sterile stamen, that is, without an anther or at least not producing pollen.

Stem end: The basal portion of an inferior ovary at fruiting time, the point of attachment of the pedicel.

Stigma: The apical portion of a pistil ie. the portion receptive to pollen. It is covered commonly with minute papillae, and it is sticky through production of a sugar solution in which the pollen grains germinate.

Stock or rootstock: It is the lower portion of the graft which develops into the root system of the grafted plants.

Style: the tubular upper or middle part of a pistil connecting the stigma and the ovary.

Subclass: A taxon of a rank between class and order.

Subculture: See "passage". With plant cultures this is process by which the tissue or explant is first subdivided, then transferred into fresh culture medium.

Subspecies: A taxon of a rank between species and variety; a group of related varieties.

Substrain: A substrain can be derived from a strain by isolating a single cell or groups of cells having properties or markers not shared by all cells of the parent strain.

Sucker: A rhizome or a branch of a rhizome which comes to the surface and grows into a leafy shoot.

Summer fallow: Allowing land to lie idle for a season to conserve water for use the following season.

Sunburn: A killing of bark near the ground surface due to reflected heat rays, a form of sun scald.

Suspension culture: A type of culture in which cells, or aggregates of cells, multiply while suspended in liquid medium.

Sterility: The inability to set and mature fruit with viable seed. This failure may be due to non-function of the pollen, the ovule or both.

Syconium: An enlarged hollow receptacle bearing flowers and ultimately fruits inside.s

Sympetalous: All the petals coalescent at least basally.

Syncarp: A structure composed of several fruits, these more or less coalescent as for example the fruits of pineapple or mulberry.

Syncarpous: With coalescent carpels.

Syngenesious: With anthers cohering in a circle.

Synsepalous: The sepals coalescent.

Taxon: A category used in classification as for example, a variety, a species, genus, family etc. Pural. taxa.

Taxonomy: The principles of classification.

Tepal: A term used for sepals and petals which are similar and not easily distinguished from each other.

Terminal bud: Bud at the end of the stem or a branch.

Testa: The seedcoat, that is, the hardened mature integument.

Tetradynamous: Having four long and two short stamens. A condition found almost throughout the mustard family.

Tetragynous: With four pistils.

Throat: The opening of a sympetalous corolla or a synsepalous calyx, that is, the expanding part between the proper tube and the limb.

Thyrse: A densely congested panicle, the main axis indeterminate, the lateral typically determinate and therefore cymose. A mixed inflorescence.

Tissue: Groups of cells having common functions.

Tissue culture: The term used for range of procedures used to maintain and grow plant tissues and organs in aseptic culture.

Totipotency: A cell characteristic in which the potential for forming all the cell types in the adult organism is retained.

Transpiration: The phenomenon of giving off water in form of vapours.

Tuber: A thickened short underground branch of the stem serving as a storage organ containing reserve food. ex potato.

Umbel: An inflorescence with the pedicles of the flower arising from approximately the same point. A compound umbel is umbal of umbels.

Unfruitful: A plant or variety that fails to set a commercial crop of fruit and mature it.

Unisexual: Of only one sex. Descriptive of a flower having only stamens or only pistils, not both, or of a gymnosperm producing only pollen or only ovules or of the gametophytes of pteridophyte bearing the reproductive organs of only one sex.

Utricle: A small, seeded more or less indehiscent fruit which appears to be inflated, that is, with a relatively thin pericarp more or less remote from the single seed. At maturity the utricle opens either irregularly or along a horizontal line.

Vascular cambium: It is a thin tissue located between the bark (periderm, cortex and phloem) and the wood (xylem).

Vermiculite: It is a micaceous mineral that expands markedly when heated. It is very light in weight, neutral in reaction with good buffering properties and insoluble in water.

Vivipary: The phenomenon in which seeds germinate in the fruit while still attached to the plant.

Water berries: A disorder of the grape in which the fruits are watery and fails to ripen properly.

Wind burn: A disorder of the leaves in which first their edges and later perhaps the entire leaf dries out and present a scorched appearance.

Water spot: A disorder of orange fruits characterized by water soaked appearance.

Xenia: The direct influence of foreign pollen on the part of the mother plant that develops into endosperm.

Xyloporosis: A disorder associated with lack of compatibility between stock and scion, characterized by pores or pits in the wood and corresponding pegs in the bark.

Zygomorphic: Bilaterally symmetrical. Often this is described as irregular, especially with reference to corollas.

Zygote: The fertilized egg, resulting from combining of the male and female gametes.

Bibiliography

Arora, J.S. 2006. Introductory Ornamental Horticulture. Kalyani Publisher, Delhi, pp. 67-76.

Baily, L.H. 1963. The standard cyclopedia of horticulture, Macmillan and Co., New York.

Benito, M., Masaguer, A., De Antonio, R. and Moliner, A. 2005. Use of pruning compost as a component in soilless growing media. *Biosource Technology*, 96: 597-603.

Bhalla, R., Sharma, S., Dhiman, S.R. and Ritu, J. 2006. Effect of biofertilizers and biostimulant on growth and flowering in standard carnation (*Dianthus caryophyllus* Linn.). *J. Ornam. Hortic.* 9(4): 282-285.

Bhattacharyya, H.K. 1997. Floriculture management. *Floriculture Today*, 2. 15-23.

Bose, T.K. and Yadav, L.P. 2012. Commercial Flowers, Anthurium, *Naya Prakash*, pp. 643-650.

Bumess, A. 2008. Anthurium: A Manual for Small Holder Production in Fiji, *South Sea Orchids*, CTA.

Chadha, K.L. and Bhattacharjee, S.K. 1995. Advances in Horticulture, Vol. 12 Ornamental Plants, Malhotra Publishers, New Delhi (India).

Chandler, F. 1991. Commercial Anthurium Production, Factsheet, Caribbean Agricultural Research and Development Institute.

Cohen, D. 1982. Title comb. Proc. Intern. Pl. Prop. Soc., 31: pp 312-6.

Dalal, S.R., Wankar, A.M. and Somavanshi, A.V. 2009. Performance of carnation cultivars under polyhouse condition. *The Asian Journal of Horticulture*, 4. 225-226.

DeBoodt, M. and O. Verdonck. 1972. The physical properties of the substrates in horticulture. *Acta Hort.*, 26:37-44.

Ebrahimzadeh, A., Jimenez-Becker, S., Manzano, S.M., Jamilena-Quesada, M. and Lao-Arenas. M.T. 2011. Evaluation of ethylene production by ten Mediterranean carnation cultivars and their response to ethylene exposure. *Span J. Agric Sci.*, 9: 524-530.

Fonteno, W.C. 1996. Growing Media: types and physical/chemical properties. P. 93-11.

Gill, A.P.S. and Arora, J.S. 1988. Performance of Sim carnation under sub-tropical conditions of Punjab. *Indian Journal of Horticulture*, 45: 329-35.

Hansen, Hans, V.A. taxonomic revision of the genus Gerbera (Compositae, Mutisieae) sections Gerbera, Parva, Piloselloides (in Africa), and Lasiopus (Opera botanica. No. 78: 1985). *ISBM 87-88702-04-9.*

Higaki, T., Lichty, J.S. and Moniz, D. 1995. Anthurium culture in Hawaii, College of Tropical Agriculture and Human Resources, Research Extension Series 152.

Kim, I.Y., Cho, H.K. 2002. Development of Standard Analysis Methods for Physical Properties on Korean bedsoil 1. Particle density and Bulk density. *Korean Journal of Soil Science and Fertilizer*. 35(6): 327-335.

Kim, I.Y., Jung, K.H. and Ro, H.M. 2002. Development of Standard Analysis Methods for Physical Properties on Korean soilless medium 2. Water content, Water retention, Saturated hydraulic conductivity. *Korean Journal of Soil Science and Fertilizer*, 35(6): 335-343.

Lalnunmawia, F. and Khawlhring, N. 2011. Cultivation of Anthurium in Mizoram, India: present scenario and future prospect. *Science Vision*, 11 (4): 203-207.

Misra, Kaushal Kumar 2011. Ornamental Gardening in India. Biotech Book, New Delhi.

Mukherjee, D. 1991. Protected cultivation for increased production of quality plants and flowers. *Floriculture Technology*, Trade and Trends. Oxford and IBH Pub. Co., Bombay. Pp. 443-451.

Mukund, S., Shirol, A.M., Reddy, B.S. and Kulkarni, B.S. 2004. Performance of standard carnation (*Dianthus caryophyllus* L.) cultivars under protected cultivation for vegetative characters. *Journal* of *Ornamental Horticulture*, 7: 212-216.

Mysore, S., Gajanana, T.M. and Dakshinamoorthy, V. 2005. Economic feasibility and profitability of carnation cultivation. *Journal* of *Ornamental Horticulture*, 8: 254-259.

Nesom, G.L. 2004. Response to "The Gerbera complex (Asteraceae, Mutisieae): to split or not to split" by Liliana Katinas. *Sida*, 21: 941-942.

Nukui, H., S., Kudo, A., Yamashita and Satohs. 2004. Repressed ethylene production in the gynoecium of long-lasting flowers of the carnation 'White Candle': role of gynoecium in carnation flower senescence. *J. Exp. Bot.*, 55: 641-650.

Ramesh, K., Singh, K. and Reddy, B.S. 2002. Effect of planting time, Photoperiod, GA_3 and pinching on carnation. *J. Ornam. Hortic.*, 5: 20-23.

Randhawa, G.S. 1973. Ornamental Horticulture in India. Today and Tomorrow Printers and Publishers, New Delhi.

Randhawa, G.S. and Mukhopadhyay, A. 1986. Floriculture in India. Allied publishers Pvt. Limited, New Delhi.

Reddy, B.S., Patil, J.R.T. and Kulkarni, B.S. 2004. Studies on vegetative growth, flower yield and quality of standard carnation (*Dianthus caryophyillus* L.) under low cost polyhouse condition. *J. Ornam. Hortic.*, 7(3-4): 217-220.

Shahri, W. and Tahir, I. 2011. Flower senescence-strategies and some associated events. *Bot Rev.*, 77:152-184.

Singh, K.D. and Sangama. 2003. Evaluation of post harvest quality of the some cultivars of carnations flowers grown in greenhouse. *J. Ornamental Hortic.*, 6(3):274-276.

Surassawadee Promyou, Ketsab, S. and Doorn, W.G.V. 2012. Salicylic acid alleviates chilling injury in anthurium (*Anthurium andraeanum* L.) flowers, *Postharvest Biology and Technology*, 64: 104-110.

Swarup, V. 1984. Export of ornamentals: Problem and prospects. *Indian Horticulture*, 29(2): 53-56.

Tija, B.O. and Funnell, K.A. 1986. Acta Hort., 181: 451-8.

Tyagi, Y.Y, Mantur, S.M. and Reddy, B.S. 2007. Effect of pinching on growth yield and quality of flower of carnation varieties grown under polyhouse. *Karnataka J. Agric. Sci.*, 20(4):816-818.

Warren, A. 2012. Zantedeschia production. *Zantedeschia Technical Bulletin*. C01/12: pp.

Woodson, W.R., Park, K.Y., drory, A., Larsen, P.B. and Wang, H., 1992. Expression of ethylene biosynthetic pathway transcripts in senescing carnation flowers. *Plant Physiol.*, 99: 526-532.

Index

N

O

P

T

U

V

W

Y

Z